Advances in Hydrometallurgy

Advances in Hydrometallurgy

Special Issue Editor
Alexandre Chagnes

MDPI • Basel • Beijing • Wuhan • Barcelona • Belgrade • Manchester • Tokyo • Cluj • Tianjin

Special Issue Editor
Alexandre Chagnes
University of Lorraine
France

Editorial Office
MDPI
St. Alban-Anlage 66
4052 Basel, Switzerland

This is a reprint of articles from the Special Issue published online in the open access journal *Metals* (ISSN 2075-4701) (available at: https://www.mdpi.com/journal/metals/special_issues/advances_hydrometallurgy).

For citation purposes, cite each article independently as indicated on the article page online and as indicated below:

LastName, A.A.; LastName, B.B.; LastName, C.C. Article Title. *Journal Name* **Year**, *Article Number*, Page Range.

ISBN 978-3-03928-939-4 (Pbk)
ISBN 978-3-03928-940-0 (PDF)

Cover image courtesy of Wikimedia.

© 2020 by the authors. Articles in this book are Open Access and distributed under the Creative Commons Attribution (CC BY) license, which allows users to download, copy and build upon published articles, as long as the author and publisher are properly credited, which ensures maximum dissemination and a wider impact of our publications.

The book as a whole is distributed by MDPI under the terms and conditions of the Creative Commons license CC BY-NC-ND.

Contents

About the Special Issue Editor vii

Preface to "Advances in Hydrometallurgy" ix

Alexandre Chagnes
Advances in Hydrometallurgy
Reprinted from: *Metals* 2019, *9*, 211, doi:10.3390/met9020211 1

Udit Surya Mohanty, Lotta Rintala, Petteri Halli, Pekka Taskinen and Mari Lundström
Hydrometallurgical Approach for Leaching of Metals from Copper Rich Side Stream Originating from Base Metal Production
Reprinted from: *Metals* 2018, *8*, 40, doi:10.3390/met8010040 3

Luis Beiza, Víctor Quezada, Evelyn Melo and Gonzalo Valenzuela
Electrochemical Behaviour of Chalcopyrite in Chloride Solutions
Reprinted from: *Metals* 2019, *9*, 67, doi:10.3390/met9010067 15

Gözde Alkan, Claudia Schier, Lars Gronen, Srecko Stopic and Bernd Friedrich
A Mineralogical Assessment on Residues after Acidic Leaching of Bauxite Residue (Red Mud) for Titanium Recovery
Reprinted from: *Metals* 2017, *7*, 458, doi:10.3390/met7110458 27

Alexandre Chagnes and Gérard Cote
Chemical Degradation of a Mixture of tri-*n*-Octylamine and 1-Tridecanol in the Presence of Chromium(VI) in Acidic Sulfate Media
Reprinted from: *Metals* 2018, *8*, 57, doi:10.3390/met8010057 39

Nathalie Leclerc, Sophie Legeai, Maxime Balva, Claire Hazotte, Julien Comel, François Lapicque, Emmanuel Billy and Eric Meux
Recovery of Metals from Secondary Raw Materials by Coupled Electroleaching and Electrodeposition in Aqueous or Ionic Liquid Media
Reprinted from: *Metals* 2018, *8*, 556, doi:10.3390/met8070556 49

Pape Diaba Diabate, Laurent Dupont, Stéphanie Boudesocque and Aminou Mohamadou
Novel Task Specific Ionic Liquids to Remove Heavy Metals from Aqueous Effluents
Reprinted from: *Metals* 2018, *8*, 412, doi:10.3390/met8060412 67

Pia Sinisalo and Mari Lundström
Refining Approaches in the Platinum Group Metal Processing Value Chain—A Review
Reprinted from: *Metals* 2018, *8*, 203, doi:10.3390/met8040203 83

Zhonglin Dong, Tao Jiang, Bin Xu, Yongbin Yang and Qian Li
Recovery of Gold from Pregnant Thiosulfate Solutions by the Resin Adsorption Technique
Reprinted from: *Metals* 2017, *7*, 555, doi:10.3390/met7120555 95

Qiang Zhong, Yongbin Yang, Lijuan Chen, Qian Li, Bin Xu and Tao Jiang
Intensification Behavior of Mercury Ions on Gold Cyanide Leaching
Reprinted from: *Metals* 2018, *8*, 80, doi:10.3390/met8010080 113

Xavier Hérès, Vincent Blet, Patricia Di Natale, Abla Ouaattou, Hamid Mazouz, Driss Dhiba and Frederic Cuer
Selective Extraction of Rare Earth Elements from Phosphoric Acid by Ion Exchange Resins
Reprinted from: *Metals* **2018**, *8*, 682, doi:10.3390/met8090682 127

Bengi Yagmurlu, Carsten Dittrich and Bernd Friedrich
Effect of Aqueous Media on the Recovery of Scandium by Selective Precipitation
Reprinted from: *Metals* **2018**, *8*, 314, doi:10.3390/met8050314 145

Ernesto de la Torre, Estefanía Vargas, César Ron and Sebastián Gámez
Europium, Yttrium, and Indium Recovery from Electronic Wastes
Reprinted from: *Metals* **2018**, *8*, 777, doi:10.3390/met8100777 159

About the Special Issue Editor

Alexandre Chagnes was awarded his Ph.D. from University of Tours, France (Physical Chemistry), in September 2002 for his thesis entitled "Thermodynamic and Electrochemical Study of Organic Electrolytes for Lithium-Ion Batteries", M.S. (June 1999) from University of Poitiers, France, for his thesis "Catalysis, Energy and Clean-Up Processes, and B.S. from University of Tours, France (Chemistry), in June 1997. Alexandre Chagnes is now Full Professor at Université de Lorraine in France. He has published 117 literature articles, 5 books, 8 book chapters, and 2 patents on various topics in solution chemistry, thermodynamic, electrochemistry, and separation science and has given 127 talks at national and international meetings. He is Director of Industrial Partnerships of the Engineering School of Geology in Nancy, the Scientific Director of Labex Ressources21, and Head of the National Research Network PROMETHEE on hydrometallurgical processes. His research focus includes hydrometallurgy, solution chemistry, nuclear chemistry, separation science, thermodynamic, electrochemistry, and lithium batteries.

Preface to "Advances in Hydrometallurgy"

The development of new technologies and the increasing demand for mineral resources from emerging countries are responsible for significant tensions in the pricing of non-ferrous metals. Some metals have become strategic and critical because they are used in many technological applications and their availability remains limited. In addition to energetic raw materials, such as oil or gas, the industry uses about 50 different metals. For many of them, the worldwide annual consumption ranges from a few tens of tons to several hundred thousand tons. Some of them, the strategic metals, are crucial for achieving high performance. They are found in high-tech products, such as flat panel TVs (indium), solar panel cells (indium), lithium-ion batteries for electric vehicles (lithium, cobalt), magnets (rare earths, such as neodymium and dysprosium), scintillators (rare earth elements), and aviation and medical applications (titanium). The secured supply of these metals is crucial to continue producing and exporting these technologies and because specific properties of these metals make them essential and difficult to substitute for a given industrial application. Hydrometallurgical processes have the advantages of being able to process low-grade ores, to allow better control of co-products, and to produce a lesser environmental impact providing that the hydrometallurgical route is optimized and cheap. With the depletion of deposits and the growing interest in low-grade elements (e.g., rare earth elements), the metallurgical industry has shown a growing interest in the development of hydrometallurgical processes more adapted to current challenges over the last 15 years. The need to develop more efficient, economical, and environmentally friendly processes, capable of extracting metals from increasingly complex and poorly polymetallic matrices, is real. The aim of this book was to highlight recent advances related to hydrometallurgy to face new challenges in metal production. Twelve contributions from experts in hydrometallurgy are published in this book, outlining recent and original advances in the fields of precious metals, processing of primary and secondary resources, and process improvement. These works seek alternative chemical technologies to extract, separate, and produce metals or metal salts. Regarding precious metals, special attention is paid to evaluating the current use and future development in gold and platinum group metal recovery. More generally, four papers were selected to introduce recent advances in the recovery of several important metals in our society: copper, which is one of the most produced base metals in the world; rare earth elements for obvious technological challenges; and scandium due to its potential application in high technologies as scientists and engineers have been working recently to develop new products incorporating this metal. Although recycling will never replace primary resources, metal extraction from spent materials and tailings should not be neglected; many challenges remain in hydrometallurgy. Four papers were selected to introduce few challenges in the recovery of rare earth elements from electronic wastes, titanium from bauxite residues, and the use of ionic liquids in the recovery of metals from wastes by liquid–liquid extraction and electrodeposition.

Alexandre Chagnes
Special Issue Editor

Editorial

Advances in Hydrometallurgy

Alexandre Chagnes

Université de Lorraine, CNRS, GeoRessources, GDR Promethee (GDR 3749), F-54000 Nancy, France; alexandre.chagnes@univ-lorraine.fr; Tel.: +33-(0)372-744-544

Received: 2 February 2019; Accepted: 6 February 2019; Published: 11 February 2019

The development of new technologies and the increasing demand of mineral resources from emerging countries are responsible for significant tensions in the price of non-ferrous metals. Some metals have become strategic and critical because they are used in many technological applications and their availability remains limited. In addition to energetic raw materials, such as oil or gas, the industry uses about fifty different metals. For many of them, the worldwide annual consumption ranges from a few tens of tons to several hundred thousand tons. Some of them, the so-called strategic metals, are crucial for achieving high performances. They are found in high-tech products, such as flat panel TVs (indium), solar panel cells (indium), lithium-ion batteries for electric vehicles (lithium, cobalt), magnets (rare earths, such as neodymium and dysprosium), scintillators (rare earths), and aviation and medical applications (titanium). The secured supply of these metals is crucial to continue producing and exporting these technologies, and because specific properties of these metals make them essential and difficult to substitute for a given industrial application.

Hydrometallurgical processes have the advantages of being able to process low-grade ores, to allow better control of co-products and to have a lower environmental impact providing that hydrometallurgical route is optimized and cheap. With the depletion of deposits and the growing interest in low-grade elements (e.g., rare earth elements), the metallurgical industry has shown a growing interest in the development of hydrometallurgical processes more adapted to current challenges over the last fifteen years. The need to develop more efficient, economical and environmentally-friendly processes, capable of extracting metals from increasingly complex and poorly polymetallic matrices, is real. The aim of this Special Issue was to highlight recent advances related to hydrometallurgy to face new challenges in metal production. For this goal, twelve papers have been published in this special issue in order to highlight interesting studies in the fields of precious metals, processing of primary and secondary resources and process improvement by understanding fundamental behavior and seeking alternative chemical technologies to extract, separate and produce metals or metal salts. Regarding precious metals, a special attention has been paid to evaluate the current use and future development in gold and platinum group metal recovery. More generally, four papers have been selected to introduce recent advances in the recovery of several important metals in our society: copper which is one of the most produced base metals in the world, rare-earth for obvious technological challenges and scandium because of its potential application in high-technologies as scientists and engineers have been working recently to develop new products incorporating this metal.

Although recycling will never replace primary resources, metal extraction from spent materials and tailings have not to be neglected and there are many challenges to face up in hydrometallurgy. Four papers have been selected to introduce few challenges in the recovery of rare earth elements from electronic wastes, titanium from bauxite residues and the use of ionic liquids in the recovery of metals from wastes by liquid-liquid extraction and electrodeposition.

Conflicts of Interest: The authors declare no conflict of interest.

 © 2019 by the author. Licensee MDPI, Basel, Switzerland. This article is an open access article distributed under the terms and conditions of the Creative Commons Attribution (CC BY) license (http://creativecommons.org/licenses/by/4.0/).

Article

Hydrometallurgical Approach for Leaching of Metals from Copper Rich Side Stream Originating from Base Metal Production

Udit Surya Mohanty [1], Lotta Rintala [2], Petteri Halli [1], Pekka Taskinen [1] and Mari Lundström [1,*]

[1] Department of Chemical and Metallurgical Engineering, School of Chemical Engineering, Aalto University, Vuorimiehentie 2, P.O. Box 16200, FI-00076 AALTO, Espoo (Otaniemi) 02150, Finland; udit.mohanty@aalto.fi (U.S.M.); petteri.halli@aalto.fi (P.H.); pekka.taskinen@aalto.fi (P.T.)
[2] VTT Technical Research Centre of Finland Ltd., Solutions for Natural Resources and Environment, Biologinkuja 7, ESPOO, P.O. Box 1000, FI-02044 VTT, Finland; lotta.rintala@vtt.fi
* Correspondence: mari.lundstrom@aalto.fi; Tel.: +358-9-47001

Received: 17 November 2017; Accepted: 5 January 2018; Published: 8 January 2018

Abstract: Pyrometallurgical metal production results in side streams, such as dusts and slags, which are carriers of metals, though commonly containing lower metal concentrations compared to the main process stream. In order to improve the circular economy of metals, selective leaching of copper from an intermediate raw material originating from primary base metal production plant was investigated. The raw material investigated was rich in Cu (12.5%), Ni (2.6%), Zn (1.6%), and Fe (23.6%) with the particle size D_{80} of 124 µm. The main compounds present were nickel ferrite ($NiFe_2O_4$), fayalite (Fe_2SiO_4), cuprite (Cu_2O), and metallic copper. Leaching was studied in 16 different solutions. The results revealed that copper phases could be dissolved with high yield (>90%) and selectivity towards nickel (Cu/Ni > 7) already at room temperature with the following solutions: 0.5 M HCl, 1.5 M HCl, 4 M NaOH, and 2 M HNO_3. A concentration of 4 M NaOH provided a superior selectivity between Cu/Ni (340) and Cu/Zn (51). In addition, 1–2 M HNO_3 and 0.5 M HCl solutions were shown to result in high Pb dissolution (>98%). Consequently, 0.5 M HCl leaching is suggested to provide a low temperature, low chemical consumption method for selective copper removal from the investigated side stream, resulting in PLS (pregnant leach solution) which is a rich in Cu and lead free residue, also rich in Ni and Fe.

Keywords: base metal production; intermediate; nickel iron oxide; fayalite; cuprite; leaching

1. Introduction

The growth in metal production has resulted in a gradual decrease in metal grades of ore deposits. Therefore, new technologies and flow-sheets are needed for the more efficient utilization of ore processing tailings, metallurgical slags, flue dusts, etc. In the base metal production, various solid side-streams are generated, such as slags, dusts, and leach residues. Inherently, these side-streams contain valuable base metals.

Thermodynamics determines the distributions of metals between metal and slag in high temperature processing [1–3]. In addition, kinetics and physical entrainment cause metal traces ending up to the slag in different steps of the production. About 60% of the world's copper and 50% of world sulphidic nickel production comes from plants using flash smelting furnace (FSF) technologies [4]. The main advantages of the FSF processes are high sulfur recovery, flexibility to feed materials and the efficient energy utilization [5]. The subsequent converting takes place in two sequential steps:

(a) The FeS elimination or slag making stage

$$2FeS_{(s)} + 3O_{2(g)} + 2SiO_{2(s)} = 2FeO \cdot SiO_{2(s)} + 2SO_{2(g)} \qquad (1)$$

(b) The copper making stage

$$Cu_2S_{(s)} + 2O_{2(g)} = 2Cu\ (s) + 2SO_{2(g)} \qquad (2)$$

As the process throughputs are generally high [6–8] the slags of the primary production can present a valuable secondary raw materials for metal recovery in future.

The composition of slags in base metal processing vary depending on the process and raw material. Copper flash smelting furnace slag generally consist of 30–50% Fe, 30–40% SiO_2, 1–10% Al_2O_3, 1–16% CaO and 0.2–1.2% of Cu [9]. Copper is mainly entrapped in the slag as chalcocite and metallic copper, as well as trace copper oxide [10]. The converter slag is usually characterized by 20–25% SiO_2, 40–45% Fe, and 5% Cu. The slags of anode furnace differs from the converter slags due its very high copper content, containing typically above 50 wt. % CuO_x, 30–35 wt. % FeO, 5–15 wt. % SiO_2, and minor amounts of As, Sb, and Pb [11,12]. Nickel flash smelting furnace slag has been reported to contain 8.7% Fe_2SiO_4, 10% Fe_3O_4, 20.5% SiO_2, 3.1% Al_2O_3, 1.3% MgO, and 1.1% CaO [13]. Generally, the slag former used is SiO_2.

Industrial smelting and converting slags are cleaned before discarding them. In most cases an electric furnace settling or reduction is used, but some copper smelters use milling and slag flotation.

In the literature, new methods for slag cleaning have been studied for eliminating trace element or cutting their internal circulations in the smelter. Thus, the impurity levels in the slags and anode copper will be lowered. Roasting of the converter slag with ferric sulphate and selective sulphation roasting are the documented pyrometallurgical methods used for the recovery of nickel, copper and zinc [14,15]. Also, pyro-hydrometallurgical methods involving acid roasting or thermal decomposition followed by water leaching have been suggested [16–18]. Various hydrometallurgical methods have been developed using lixiviants such as acids, bases, and salts for base metal extraction. Atmospheric leaching of different slag fractions has been studied in H_2SO_4, $FeSO_4$, $(NH_4)_2SO_4$, FeS_2, NaCl, and $FeCl_2$ media [19–23]. In addition, pressure leaching of copper slag containing 4.03% Cu, 0.48% Co, and 1.98% Ni at 130 °C have resulted in significant recoveries of Cu, Co, and Ni, amounting to 90% [24]. Leaching with aqueous sulfur dioxide has also proven effective in recovering 77% Co and 35% Ni from a nickel smelter slag [25].

The current study was undertaken to investigate the dissolution behaviour of selected metals, from the Cu, Ni, Fe, and Zn rich intermediate of base metal production. The focus was to dissolve copper selectively in order to produce PLS rich in copper and a residue with Fe and Ni, applicable for recovery of metals. The lixiviants used in the present study were 0.5–0.5 M HCl, 0.5–3.06 M H_2SO_4, 1–2 M HNO_3, 0.5 M NaCl + 0.1 M $CuCl_2$, 4.5 M NaCl + 0.5 M $CuCl_2$, 4.5 M NaCl + 0.1 M $CuCl_2$, and 4 M NaOH.

2. Materials and Methods

Characterization studies by Scanning Electron Microscopy (SEM), X-ray diffraction (XRD), and Particle Size Distribution (PSD) were conducted to determine the morphology, mineralogical composition, and elemental distribution of the raw material.

2.1. The Raw Material

Chemical analysis of the raw material was performed by employing microwave-assisted digestion in aqua regia (ETHOS Touch Control, Milestone Microwave Laboratory Systems, Sorisole, Italy), as aqua regia is one of the strongest and effective solvent used for metal digestion [26], Table 1. The solution analyses were conducted by ICP-OES (Inductively Coupled Plasma Optical Emission Spectroscopy, Perkin Elmer Optima 7100 DV, Waltham, MA, USA) by Milomatic Oy.

Table 1. Chemical analysis of metals of interest in raw material investigated.

Element	Concentration [wt. %]
Cu	12.5
Fe	23.6
Ni	2.6
Al	0.5
Cr	0.1
Zn	1.6
Pb	0.1
As	0.1

The particle size of the crushed intermediate raw material was analyzed by a Mastersizer 2000 laser diffraction particle size analyzer with a Scirocco 2000 Dry Powder Feeder, both manufactured by Malvern Instruments (UK). Dispersion pressure was varied from 2.0 to 3.0 bar, vibration feed rate was 50% and measurement time was varied from 12 to 30 s. Fraunhofer diffraction model was used as an optical model. The particle size distribution of the homogenized raw material is demonstrated in the volume versus particle size diagram, Figure 1. The size distribution was observed to extend from 1.4 µm to 1905 µm. The cumulative particle size distribution revealed D_{80} value of 123 µm. The mean particle size D_{10} = 13 µm, the surface weighted mean was D_{32} = 25 µm, and the volume weighted mean D_{43} = 114 µm.

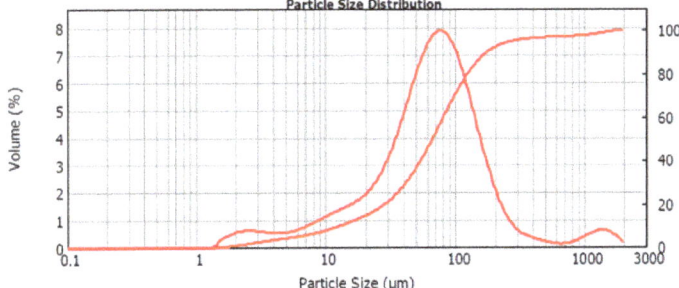

Figure 1. The observed particle size distribution of the homogenized raw material.

An X'Pert PRO-PAN Analytical X-ray diffractometer, operating at an anode current of 40 mA at 45 kV with a Cuka, by Rietveld refinement method [27] using HighScore Plus software (PANalytical), performed mineralogical analysis of the sample. Fixed Divergence Slit (FDS) 1/2° was fitted in the incident beam path to control the equatorial divergence of the incident beam and fixed incident beam. A copper mask of 15 mm was fitted in the incident beam path to control the axial width of the incident beam. Fixed Anti-Scatter Slit (FASS) 1° was used to reduce background signal. The XRD analysis of the raw material by Rietveld refinement suggested a composition of 52.2 wt. % $NiFe_2O_4$, 25.0 wt. % Fe_2SiO_4 (fayalite), 20.5 wt. % of Cu_2O (cuprite), and 2.3 wt. % of metallic Cu, Figure 2.

SEM-EDS analysis for two raw material samples was performed with a LEO 1450 VP (Carl Zeiss, Oberkochen, Germany) scanning electron microscope (SEM) and a X-MAX-50 mm² energy dispersive X-ray spectrometer (EDS) with INCA Software (Oxford Instruments, Abingdon, UK). Tungsten filament was used as a cathode and the acceleration voltage used was 15 kV.

The raw material samples were cast in epoxy and treated in vacuum, for eliminating gas bubbles attached into the particles, and prepared for SEM-EDS examination polished sections using standard wet methods. It can be discerned from Figure 3 that a larger particle of size around 500 µm is encompassed by smaller particles of particle size ranging 2–50 µm in the raw material. Three phases could be observed, one of the larger particles and two phases in the smaller particles. The average weight percentages of the elements detected in spectrum 1–14 in Figure 3 are presented in Table 2. The lightest color in the back scattered electron (BSE) image corresponds to the phase of the larger particle. It consisted of an average of 88.6 wt. % Cu, 1.9 wt. % Fe, 8.8 wt. % O, and 0.6 wt. % Si (Spectra 1–5), suggesting the presence of Cu and Cu_2O (cuprite), as analyzed oxygen eventually is trace from a surface contamination. The light-gray areas in spectra 6, 9, and 12 correspond to an average of 2.2 wt. % Cu, 52.4 wt. % Fe, 14.7 wt. % Ni, 23.8 wt. % O, and 0.6 wt. % Si (Spectra 6–9, 14), indicating the three main phases, namely Fe_2SiO_4, possibly $NiFe_2O_4$ and Cu_2O. Nevertheless, the dark-gray region represented by Spectra 10–12, consisted of an average of 3.2 wt. % Cu, 3.2 wt. % Fe, 1.1 wt. % Ni, 45.8 wt. % O, and 32.3 wt. % Si, corresponding to the presence of almost pure SiO_2. Spectra 13 corresponds to epoxy, where samples were casted.

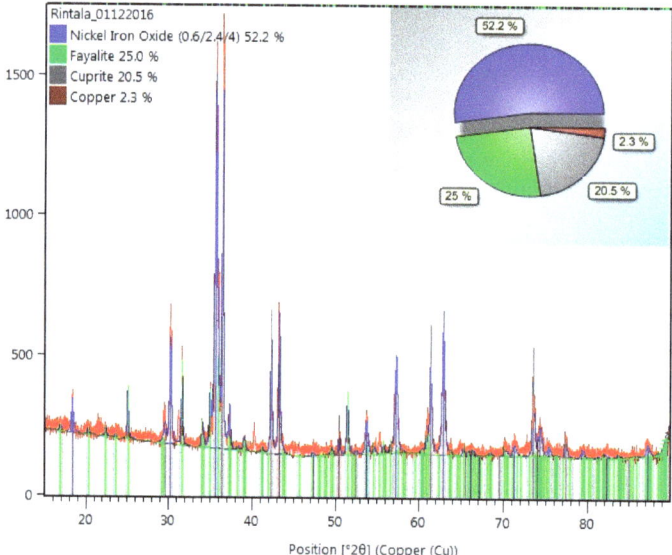

Figure 2. The obtained X-ray diffraction (XRD) pattern of the raw material.

Table 2. SEM-EDS point analysis of the particles presented in Figure 3.

[wt. %]	Spectra #1–5	Spectra #6–9, 14	Spectra #10–12
Cu	88.6	2.2	3.2
Fe	1.9	52.4	3.2
Ni	-	14.7	1.1
O	8.8	23.8	45.8
Si	0.6	0.6	32.3
Na	-	-	0.9
Mg	-	1.0	3.7
Al	-	1.7	3.6
K	-	-	1.9
Ca	-	-	0.7
Ti	-	0.8	0.4
Cr	-	1.9	-
Zn	-	2.0	-
Pb	-	-	4.3

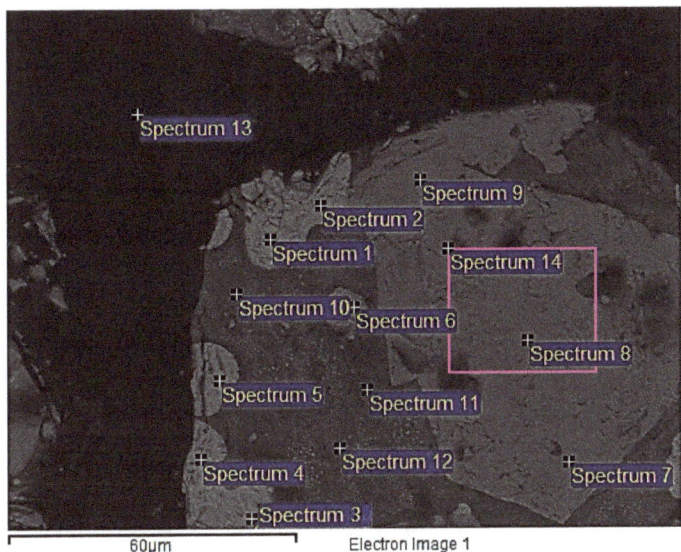

Figure 3. Back scattered Scanning Electron Microscopy (SEM) micrograph of the overall raw material. Spectra 1–5 (Cu_2O phase), Spectra 6–9, 14 ($NiFe_2O_4$ phase).

2.2. Leaching Experiments

In order to investigate the extraction without external heating, leaching was conducted at ambient temperature (25 °C) for 48 h in several solutions (Table 3). Leaching experiments were conducted in an Erlenmeyer flasks and the solutions were mixed by an IKA RO 10 Multi Station Digital Magnetic Stirrer at 300 RPM. The used S/L ratio was 0.025 (5 g solids/200 mL solution). To evaluate the leaching efficiency of Ni, Zn, Cr, Pb, Cu, Fe, and Al, the solution was filtered after the leaching step and the filtrate was analysed by AAS (atomic absorption spectrophotometer), using a Varian AA240 (Varian, Palo Alto, CA, USA), and ICP-OES [28,29].

Table 3. Solutions used in the leaching tests.

Solution	Concentrations	Chemicals	Manufacturer (Grade)
HCl	0.5 M 1.5 M 2.5 M 3.0 M 5 M	HCl 37%	EMPARTA ACS (for analysis)
H_2SO_4	0.51 M 1.22 M 1.93 M 2.65 M 3.06 M	H_2SO_4 95–97%	EMSURE ISO (for analysis)
HNO_3	1 M	HNO_3 65%	EMSURE (for analysis)
$CuCl_2$, pH 1	0.5 M NaCl + 0.1 M $CuCl_2$ 4.5 M NaCl + 0.5 M $CuCl_2$ 4.5 M NaCl + 0.1 M $CuCl_2$	$CuCl_2 \cdot 2H_2O$	VWR Chemicals (technical)
NaOH	4 M	NaOH	SIGMA-ALDRICH (technical)

No external oxidation by gas bubbling was used in the experiments. Redox potential was measured by a Fluke 115 True RMS Multimeter using platinum wire and Saturated Calomel Electrode (SCE). Mettler Toledo Seven (Easy pH meter) was used for pH measurements, except in the NaOH solutions, where Hanna Instruments Edge pH meter was employed.

3. Results and Discussion

Leaching was performed on the raw material to get an insight into the dissolution phenomena related to Cu, Ni, Zn, and Fe in various lixiviants. Also leaching of trace metals, such as Cr, Pb, and Al, was explored. The aim was to find a selective, low temperature, and low chemical consumption leaching procedure for copper present in the raw material. Furthermore, the target was to leave nickel in the leach residue in the leaching stage.

Table 4 presents the metal yields to the solution in all 16 investigated media. The corresponding redox potentials, as well as pHs before and after the experiment are presented in Figure 4. It can be seen that there is some variety in the recovery percentage—this is most likely attributed to the heterogeneous nature of the investigated raw material with big particle size and wide particle size range combined with small solid/liquid ratio in the leaching experiments. This leads in to some variation in the representativeness of each sample, thus also resulting some error in the recovery calculations.

Table 4. Extraction of investigated metals from the raw material (%).

Solution	Ni	Cu	Fe	Zn	Cr	Pb	Al
0.5 M HCl	10	*	55	40	20	98	39
1.5 M HCl	18	*	78	48	45	93	56
2.5 M HCl	43	95	81	64	67	97	69
3 M HCl	97	72	54	66	84	99	71
5 M HCl	96	86	74	92	55	97	79
0.5 M H_2SO_4	35	70	53	63	84	21	51
1.22 M H_2SO_4	64	77	60	76	56	23	63
1.93 M H_2SO_4	77	81	82	80	62	23	69
2.65 M H_2SO_4	86	71	78	86	57	23	65
3.0 M H_2SO_4	81	65	62	86	56	17	65
4.5 M NaCl + 0.5 M Cu^{2+} pH 1	1	5	-	1	-	65	7
4.5 M NaCl + 0.1 M Cu^{2+} pH 1	1	3	-	1	-	62	5
0.5 M NaCl + 0.1 M Cu^{2+} pH 1	3	61	-	27	-	53	5
1 M HNO_3	3	79	-	30	-	98	16
2 M HNO_3	4	93	-	30	-	*	13
4 M NaOH	0.3	*	-	2	-	59	22

* Full leaching.

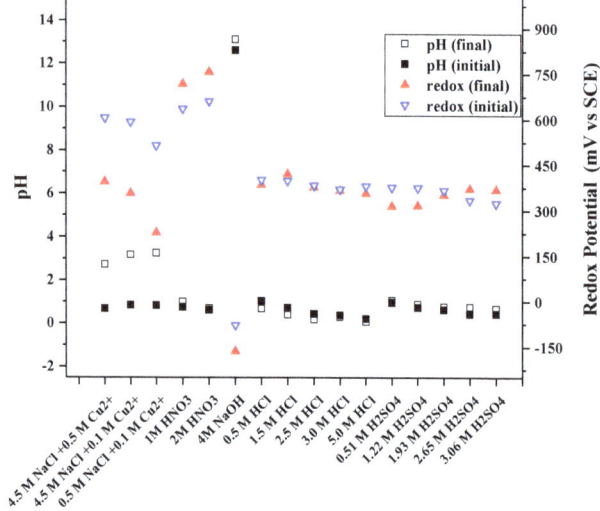

Figure 4. Measured redox potentials and pH during leaching in 16 investigated leaching media.

3.1. Leaching of Copper

Table 4 shows that copper was dissolved well into most lixiviants investigated. The highest extraction of Cu was achieved with 1.5 M HCl. Also 4 M NaOH, 0.5 M HCl, 2.5 M HCl, and 2 M HNO_3 resulted in yields higher than 93%, and 1 M HNO_3, 1.22 M H_2SO_4, and 1.93 M H_2SO_4 showed >75% extraction. The chloride leaching experiments (0.1 and 0.5 M of copper (II) as oxidant along with 4.5 M NaCl) showed only minor Cu dissolution ($\leq$5%), most likely due to a final pH close to 3 (see Figure 4), indicating copper precipitation as atacamite [30]. Sulfuric acid concentration increase was shown to increase Cu extraction up to 80% at 1.93 M, however at higher concentrations the extraction was decreased, being 65% at 3.0 M H_2SO_4. The extraction efficiency of copper was found to be comparatively lower in H_2SO_4 than in HCl and HNO_3 medium (Table 4). Habashi et al. [31] have suggested that since HCl and HNO_3 generate 1 mole of H^+ ions when dissolved in water, they produce similar dissolution efficiency compared to H_2SO_4, which produces 2 moles of H^+ ions. Also, the extraction efficiency of Cu was higher in 2 M HNO_3 than in 1 M HNO_3 (Table 4) as the oxidizing potential of NO_3^- ions has been reported to increase with increase in solution acidity [32].

In chloride media, it is suggested that cuprous chloride complexes $CuCl_3^{2-}$ and $CuCl_4^{3-}$ will be produced sequentially from $CuCl_2^-$ with chloride concentration above 1 M [33]. Chloride ions complexes can stabilize Cu(I) ions thereby increasing copper solubility. The complexation also increases the redox potential of Cu(II)/Cu(I) thereby enhancing the oxidative power of the solution. Copper is also known to be dissolvable at high pHs such as in 4 M NaOH media. The pH values measured in NaOH leaching (Figure 4) suggest the prevailing species as $Cu(OH)_3^-$ [34].

The suggested reactions acid/basic leaching reactions in HCl (3), sulfuric acid (4), and basic NaOH (5) for Cu_2O, are presented below with their standard Gibbs energies of the reactions at 25 °C from HSC Chemistry database [35]:

$$Cu_2O + 8HCl_{(a)} = 2CuCl_4^{3-}{}_{(a)} + 6H^+{}_{(a)} + H_2O_{(a)}, \Delta G° = -121.28 \text{ kcal/mol} \quad (3)$$

$$Cu_2O + H_2SO_{4(l)} = Cu + CuSO_{4(ia)} + H_2O_{(a)}, \Delta G° = -14.71 \text{ kcal/mol} \quad (4)$$

$$Cu_2O + 4NaOH_{(a)} + H_2O_{(a)} = 2Cu(OH)_3^-{}_{(a)} + 4Na^+{}_{(a)} + 2e^-, \Delta G° = -28.13 \text{ kcal/mol} \quad (5)$$

The species (a), (ia) and (l) refers to aqueous, neutral aqueous and liquid phase.

3.2. Ni Leaching and Selectivity between Copper and Nickel

According to the mineralogy, the prevailing nickel phase in the raw material investigated is nickel ferrite $NiFe_2O_4$. Ferrites are known to be refractory in leaching. This is confirmed by the results which showed that the maximum Ni extraction (97%) was observed in aggressive concentrated leaching media (3–5 M HCl). The suggested leaching reactions in HCl are presented in (6) and (7). From the speciation diagram of nickel containing $NiCl_2$ and HCl [36], most nickel is suggested to exist as Ni^{2+} up to 5 M HCl. However, the concentration of $NiCl^+$ gradually increases with increases in HCl. Nickel dissolution did not show any selectivity versus iron in any of the leaching media investigated. This is due to the dominating Ni phase $NiFe_2O_4$ resulting in a simultaneous Ni and Fe dissolution. Also, in the absence of neutralization, no back precipitation was observed.

$$NiFe_2O_4 + 8HCl_{(a)} = Ni^{+2}{}_{(a)} + 2FeCl^{+2}{}_{(a)} + 4H_2O_{(a)} + 6Cl^-{}_{(a)}, \Delta G° = -127.78 \text{ kcal/mol} \quad (6)$$

$$NiFe_2O_4 + 8HCl_{(a)} = NiCl^+{}_{(a)} + 2FeCl_2^+{}_{(a)} + 4H_2O_{(a)} + 3Cl^-{}_{(a)}, \Delta G° = -161.17 \text{ kcal/mol} \quad (7)$$

The current study aims to selectively dissolve Cu versus nickel. Figure 5 presents the dissolved Cu/Ni ratio in solution with eight of the most selective lixiviants. It can be seen that the highest selectivity was achieved with 4 M NaOH (w(Cu):w(Ni) = 340 in solution). Also 1 and 2 M HNO_3 (w(Cu)/w(Ni) = 26 and 23) provided excellent selectivity as well as 0.5 M HCl solution (w(Cu)/w(Ni) = 10).

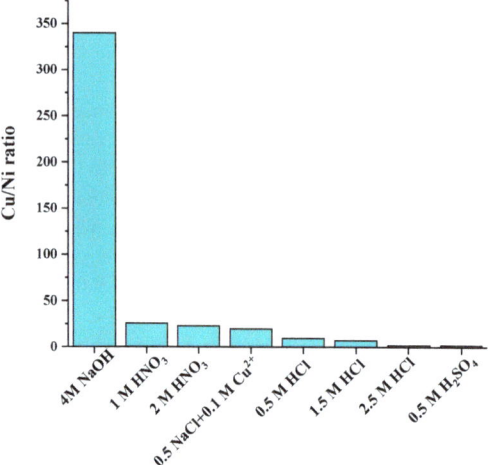

Figure 5. Cu selectivity against Ni in the best eight selective leaching media.

The lowest Ni dissolution (≤4%) was observed with 4 M NaOH, 1–2 M HNO$_3$, and all the investigated NaCl solutions. Dilute HCl (0.5 M) dissolved only 10% of nickel.

3.3. Leaching of Iron

Iron originated from the two main minerals of the raw material, NiFe$_2$O$_4$ and Fe$_2$SiO$_4$. Most notable extractions of Fe (81–82%) was observed in 1.93 M H$_2$SO$_4$ and 2.5 M HCl. Furthermore, the Fe extraction was high (>50%) in all hydrochloric and sulphuric acid media. Generally, no selectivity between iron and nickel or copper was found. However, minor selectivity between Cu and Fe was observed in 0.5 M and 1.5 M HCl (Cu/Fe = 1.9 and 1.5, respectively).

In chloride media, Fe(III) forms FeCl^{2+} and Fe^{3+} at lower Cl$^-$ concentrations, whereas FeCl$_2^+$ is formed at higher chloride concentration [37,38]. The suggested reactions for NiFe$_2$O$_4$ are presented earlier in (6) and (7), in addition fayalite is suggested to leach according to reactions (8) and (9):

$$Fe_2SiO_4 + 2H_2SO_{4\,(l)} = 2FeSO_{4\,(ia)} + H_4SiO_{4(a)}, \Delta G° = -51.67 \text{ kcal/mol} \tag{8}$$

$$H_4SiO_4 \text{ (colloid)} = SiO_2 \cdot 2H_2O \tag{9}$$

It is clear that iron dissolution is strongly related to the solution pH. At pHs < 2 iron is known to remain soluble [39]. This can be taken as an advantage in the leaching, as pH adjustment can significantly improve the selectivity between Cu and Fe.

3.4. Leaching of Zinc

The maximum extraction of Zn (90%) was achieved in 5 M HCl. Several researchers [40,41] have reported 90% recovery of Zn in leaching of zinc ferrite in the concentration range of 0.5–6 M HCl. When Zn(II) is dissolved into chloride media, it is known to form complex such as ZnCl$_3^-$ [42]. Decrease in the Zn yield from 86% to 40% was noticed in the following order of the lixiviants:

2.65 M H$_2$SO$_4$ > 3.0 M H$_2$SO$_4$ > 1.93 M H$_2$SO$_4$ > 1.23 M H$_2$SO$_4$ > 3 M HCl > 2.5 M HCl > 0.5 M H$_2$SO$_4$ > 0.5 M HCl

However, Zn extraction was ≤30% in 1–2 M HNO$_3$ and 0.5 M NaCl + 0.1 M Cu^{2+}. The alkaline leaching of zinc ferrite (4 M NaOH) was shown to result in lower Zn extraction compared to HCl and

H_2SO_4. Zn extraction was found to be 2% in 4 M NaOH. Thus, zinc ferrite fraction of magnetite was not decomposed even in strong alkaline media. The suggested reactions for Zn dissolution in HCl (10), and sulfuric acid (11) media are:

$$ZnFe_2O_4 + 8HCl_{(a)} = ZnCl_3^-{}_{(a)} + 2FeCl_2^+{}_{(a)} + 4H_2O_{(l)} + 3Cl^-{}_{(a)}, \Delta G° = -124.77 \text{ kcal/mol} \quad (10)$$

$$ZnFe_2O_4 + H_2SO_{4(l)} = ZnSO_{4\,(a)} + Fe_2O_3 + H_2O_{(l)}, \Delta G° = -26.98 \text{ kcal/mol} \quad (11)$$

3.5. Leaching of Cr, Pb, and Al

The dissolution of minor elements, such as Cr, Pb, and Al, was observed in the investigated leaching solutions. Figure 6 displays a complete 3D schematic representation of the elements of Al, Pb, and Cr and shows that generally aluminum had a high solubility into sulfuric acid and HCl media. Lead had high tendency towards HCl, NaOH, HNO_3, and chloride leaching, but it was dissolved only slightly into sulfuric acid media, evidently due to the low solubility of lead sulfate.

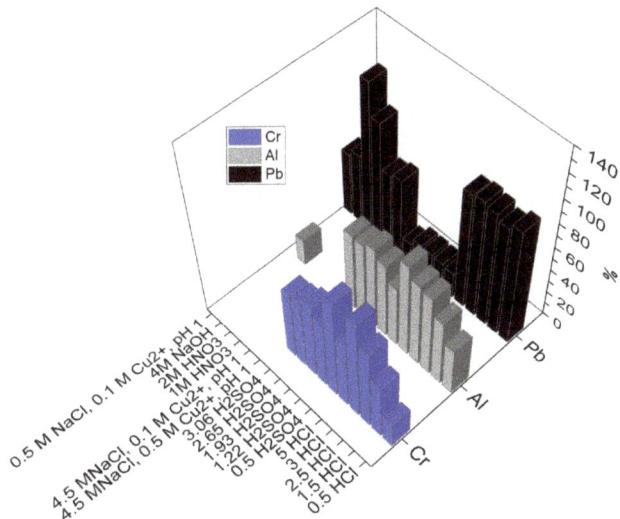

Figure 6. 3D plot displaying the yields of Al, Pb, and Cr obtained from the leaching in various lixiviants.

It was also observed from the above graph (Figure 6) that the maximum Cr yield was 83% in 3 M HCl. Furthermore, the yield of Cr decreased from 66 to 20% in presence of lixiviants in the following order: 2.5 M HCl > 1.93 M H_2SO_4 > 0.5 M H_2SO_4 > 2.65 M H_2SO_4 > 3.06 M H_2SO_4 > 1.22 M H_2SO_4 > 5 M HCl > 1.5 M HCl.

Highest aluminum extraction (>70%) was observed in 3 and 5 M HCl solutions. Additionally, all sulphuric acid solutions and HCl solutions ≥1.5 M resulted in higher than 50% leaching. Yield percentage of Pb in all the investigated HCl solutions was high (93–99%), as Pb dissolves forming chloride complexes $PbCl^+$ and $PbCl_2$ at low Cl^- concentrations, and forms $PbCl^{3-}$ and $PbCl_4^{2-}$ at higher concentrations [43].

3.6. A Comparison of the Used Lixiviants

Table 5 shows a comparison of the best copper lixiviants and the selectivities of copper dissolution for nickel and iron. It can be seen that 4 M NaOH has the highest selectivity in copper extraction. In addition, it has the highest selectivity for Zn as well. The goal of mild, selective, and low temperature leaching of copper is best approached by the 0.5 M HCl media, which was able to dissolve all copper with only 10% nickel extraction, selectivity between copper and nickel concentrations in the solution being 10. Furthermore, 0.5 M HCl was the media showing the highest selectivity towards Fe (Cu/Fe = 1.9) without pH adjustment.

Table 5. The best lixiviants selected for Cu, Fe, Pb, and Zn extraction.

Solution	Cu Extraction (%)	Cu/Ni Selectivity	Cu/Zn Selectivity	Cu/Fe Selectivity	Pb Extraction (%)	Zn Extraction (%)
1.5 M HCl	*	7	3	1.5	93	48
0.5 M HCl	*	10	3	1.9	98	40
4 M NaOH	*	340	51	-	59	2
0.5 M NaCl, 0.1 M Cu^{2+}	61	20	2	-	53	27
2 M HNO_3	93	23	3	-	*	30
1 M HNO_3	79	26	3	-	98	30

* Full leaching.

The calculated composition of the residue after 0.5 M HCl leaching suggests the leach residue composition being 23 mg/g Ni, 106 mg /g Fe, 9 mg/g Zn, 2 mg/g Al, 1.9 mg/g Cu, 0.9 mg/g Cr, and 0.02 mg/g Pb. The advantage of 0.5 M HCl is that it could dissolve also almost all Pb (98%).

4. Conclusions

Along the principles of circular economy, the recovery of metals from industrial side streams, waste, and intermediate fractions is of increasing importance. In the current study, the leaching phenomena and selective leaching of copper was investigated from an intermediate raw material originating from base metal production, with a mineralogy of 52.2% $NiFe_2O_4$, 25.0% Fe_2SiO_4 (fayalite), 20.5% of Cu_2O (cuprite), and 2.3% of metallic Cu. In the raw material, the large particles were shown to consist mainly of Cu_2O and elemental Cu.

Copper present in the raw material was shown to be easily dissolvable, over 98% Cu could be dissolved with 0.5 M, 1.5 M HCl, and 4 M NaOH. In addition, 0.5 M HCl was shown to provide selectivity towards Ni, with the Cu/Ni concentration ratio in solution being 10. Alkaline leaching in 4 M NaOH resulted in the highest selectivity for copper leaching, with the ratio of dissolved elements of Cu/Ni = 340 and Cu/Zn = 51. Also 1 and 2 M HNO_3 provided high selectivity for copper dissolution with a Cu/Ni ratio of 26 and 23, respectively. Aluminum showed high dissolution into sulfuric acid and hydrochloric acid media, the highest Al extraction being 79% in 5 M HCl whereas lead dissolution was strong in HCl, chloride, NaOH, and HNO_3. The highest extraction for Ni was obtained in 5 M HCl.

The results indicate that from the 16 investigated leaching media, hydrochloric acid leaching (0.5 M HCl) presents the lowest concentration solution matrix for selective and high copper extraction, even at room temperature.

Acknowledgments: This study is a part of NewEco project of EIT Knowledge and Innovation Community Raw Materials consortium. The authors express their deep gratitude to Petri Latostenmaa from Boliden Harjavalta for providing the raw material. The authors also acknowledge Hannu Revitzer for performing the ICP and chemical analyses and Janne Vuori for performing the particle size analyses. Also METSEK project funded by Association of Finnish Steel and Metal Producers, and Raw MATERS Finland Infrastructure supported by Academy of Finland is greatly acknowledged.

Author Contributions: Mari Lundström and Pekka Taskinen conceived and designed the experiments; Lotta Rintala performed the experiments; Petteri Halli analyzed the data; Udit Surya Mohanty and Mari Lundström wrote the paper.

Conflicts of Interest: The authors declare no conflict of interest.

References

1. Avarmaa, K.; O' Brien, H.; Johto, H.; Taskinen, P. Equilibrium distribution of precious metals between slag and copper matte at 1250–1350 °C. *J. Sustain. Metall.* **2015**, *1*, 216–228. [CrossRef]

2. Avarmaa, K.; Johto, H.; Taskinen, P. Distribution of precious metals (Ag, Au, Pd, Pt, and Rh) between copper matte and iron silicate slag. *Metall. Mater. Trans. B* **2016**, *47*, 244–255. [CrossRef]
3. Avarmaa, K.; O'Brien, H.; Taskinen, P. Equilibria of Gold and Silver between Molten Copper and FeO$_x$-SiO$_2$-Al$_2$O$_3$ Slag in WEEE Smelting at 1300 °C. In *Advances in Molten Slags, Fluxes, and Salts: Proceedings of the 10th International Conference on Molten Slags, Fluxes and Salts*; John Wiley & Sons, Inc.: Hoboken, NJ, USA, 2016; pp. 193–202.
4. Jyrkonen, S.; Haavanlammi, K.; Luomala, M.; Karonen, J.; Suikkanen, P. Processing of PGM containing Ni/Cu bulk concentrates in a sustainable way by Outotec Direct Nickel Flash Smelting process. In *Ni-Co 2013*; Springer: Cham, Switzerland, 2013; pp. 325–334.
5. Taskinen, P. Direct-to-blister smelting of copper concentrates: The slag fluxing chemistry. *Miner. Process. Extr. Metall.* **2011**, *120*, 240–246. [CrossRef]
6. Taskinen, P.; Seppala, K.; Laulumaa, J.; Poijarvi, J. Oxygen pressure in the Outokumpu flash smelting furnace—Part 1: Copper flash smelting settler. *Miner. Process. Extr. Metall.* **2001**, *110*, 94–100. [CrossRef]
7. Davenport, W.G.; King, M.J.; Schlesinger, M.E.; Biswas, A.K. *Extractive Metallurgy of Copper*, 4th ed.; Elsevier: Amsterdam, The Netherlands, 2002; p. 452.
8. Taskinen, P.; Dinsdale, A.; Gisby, J. Industrial slag chemistry: A case study of computational thermodynamics. *Scand. J. Metall.* **2005**, *34*, 100–107. [CrossRef]
9. Mihailova, I.; Mehandjiev, D. Characterisation of fayalite from complexes. *J. Chem. Technol. Metall.* **2010**, *45*, 317–326.
10. Deng, T.; Ling, Y.H. Chemical and mineralogical characterisations of a copper converter slag. *Rare Met.* **2002**, *21*, 175–178.
11. Petkov, V.; Jones, P.T.; Boydens, E.; Blanpain, B.; Wollants, P. Chemical corrosion mechanisms of magnesia–chromite and chrome-free refractory bricks by copper metal and anode slag. *J. Eur. Ceram. Soc.* **2007**, *27*, 2433–2444. [CrossRef]
12. Taskinen, P.; Kojo, I. Fluxing options in the direct-to-blister copper smelting. In Proceedings of the Molten 2009 Conference, Santiago, Chile, 18–21 January 2009; pp. 1140–1151.
13. Li, Y.; Papangelakis, G.V.; Ilya, P. High pressure oxidative acid leaching of nickel smelter slag: Characterization of feed and residue. *Hydrometallurgy* **2009**, *97*, 185–193. [CrossRef]
14. Altundogan, H.S.; Tumen, F. Metal recovery from copper converter slag by roasting with ferric sulphate. *Hydrometallurgy* **1997**, *44*, 261–267. [CrossRef]
15. Sanchez, M.; Parada, F.; Parra, R.; Marquez, F.; Jara, R.; Carrasco, J.; Palcios, J. Management of copper pyrometallurgical slags: Giving additional value to copper mining industry. In Proceedings of the VII International Conference on Molten Slags Fluxes and Salts, Cape Town, South Africa, 25–28 January 2004; pp. 543–550.
16. Geveci, A.; Topkaya, Y.; Gerceker, E. Recovery of Copper and zinc from copper converter flue dusts. In Proceedings of the 10th International Metallurgy and Material Congress, Istanbul, Turkey, 24–28 May 2000; pp. 59–68.
17. Yıldız, K.; Alp, A.; Aydın, A.O. Utilization of copper refining slags by a pyro-hydrometallurgical method. In Proceedings of the 10th International Metallurgy and Material Congress, Istanbul, Turkey, 24–28 May 2000; pp. 127–132.
18. Arslan, F.; Giray, K.; Onal, G.; Gurkan, V. Development of a Flowsheet for Recovering Copper and Tin from Copper Refining Slags. *Eur. J. Miner. Process. Environ. Prot.* **2002**, *2*, 94–102.
19. Anand, S.; Das, R.P.; Jena, P.K. Reduction—Roasting and ferric chloride leaching of copper converter slag for extracting copper, nickel and cobalt. *Hydrometallurgy* **1981**, *7*, 243–252. [CrossRef]
20. Sukla, L.B.; Panda, S.C.; Jean, P.K. Recovery of cobalt, nickel, and copper from converter slag through roasting with ammonium sulphate and sulfuric acid. *Hydrometallurgy* **1986**, *16*, 153–165. [CrossRef]
21. Tumen, F.; Bailey, N.T. Recovery of metal values from copper smelter slags by roasting with pyrite. *Hydrometallurgy* **1990**, *25*, 317–328. [CrossRef]
22. Herreros, O.; Quiroza, R.; Manzanob, E.; Bou, C.; Vinalsb, J. Copper extraction from reverberatory and flash furnace slags by chlorine leaching. *Hydrometallurgy* **1998**, *49*, 87–101. [CrossRef]
23. Tumen, F. Metal recovery from secondary copper slag by roasting with ammonium sulphate. *Turkish J. Eng. Environ. Sci.* **1994**, *18*, 1–5.

24. Anand, S.; Rao, K.S.; Jena, P.K. Pressure leaching of copper converter slag using dilute sulphuric acid for the extraction of cobalt, nickel and copper values. *Hydrometallurgy* **1983**, *10*, 305–312. [CrossRef]
25. Gbor, P.K.; Ahmed, I.B.; Jia, C.Q. Behaviour of Co and Ni during aqueous sulphur dioxide leaching of nickel slag. *Hydrometallurgy* **2000**, *57*, 13–22. [CrossRef]
26. Niemela, A.; Pitkaaho, S.; Ojala, S.; Keiski, R.L.; Peramaki, P. Microwave-assisted aqua regia digestion for determining platinum, palladium, rhodium and lead in catalyst materials. *Microchem. J.* **2012**, *101*, 75–79. [CrossRef]
27. Young, R.A. *The Rietveld Method*; Oxford University Press: Oxford, UK, 1995; 308p, ISBN 9780198559122.
28. Cardellicchio, N.; Buccolieri, A.; Di Leo, A.; Spada, L. Heavy metals in marine sediments from the MarPiccolo of Taranto (Ionian Sea, Southern Italy). *Ann. Chim.* **2006**, *96*, 727–741. [CrossRef] [PubMed]
29. Cardellicchio, N.; Buccolieri, A.; Di Leo, A.; Librando, V.; Minniti, Z.; Spada, L. Methodological approach for metal pollution evaluation in sediments collected from the Tarnto Gulf. *Toxicol. Environ. Chem.* **2009**, *91*, 1273–1290. [CrossRef]
30. Lundstrom, M.; Liipo, J.; Karonen, J.; Aromaa, J. Dissolution of six sulfide concentrates in the hydrocopper environment. In Proceedings of the South African Institute of Mining and Metallurgy Base Metals Conference, Kasane, Botswana, 27–31 July 2009; pp. 127–138.
31. Habbache, N.; Alane, N.; Djerad, S.; Tifouti, L. Leaching of copper oxide with different acid solutions. *Chem. Eng. J.* **2009**, *152*, 503–508.
32. Pacović, N.V. *Hydrometallurgy*; ŠRIF: Bor, Serbia, 1980. Chapter 3. (In Serbian)
33. Carneiro, M.F.C.; Leao, V.A. The role of sodium chloride on surface properties of chalcopyrite leached with ferric sulphate. *Hydrometallurgy* **2007**, *87*, 73–82. [CrossRef]
34. Garrels, R.M.; Thompson, M.E. Oxidation of pyrite by iron sulfate solutions. *Am. J. Sci.* **1960**, *258*, 57–67.
35. Roine, A. Sustainable Process Technology and Engineering, A Manual on HSC program, Continuous Research & Development. Outotec Research Centre: Finland, 8 March 2017.
36. Lee, M.S.; Nam, S.H. Chemical Equilibria of Nickel chloride in HCl solution at 25 °C. *Bull. Korean Chem. Soc.* **2009**, *30*, 2203–2207.
37. Ashurst, K.G. The Thermodynamics of the formation of chlorocomplexes of iron (III), cobalt (II), iron (II), manganese (II) in perchlorate medium. *Nat. Inst. Metall.* **1976**, *1820*, 1–43.
38. Peek, E.M.; Van Weert, G. Chloride Metallurgy. In Proceedings of the 32nd Annual Hydrometallurgy Meeting and International Conference of the Practice and Theory of Chloride/Metal Interaction, Montréal, QC, Canada, 19–23 October 2002; pp. 760–780.
39. Misawa, T. The thermodynamic consideration for Fe-H$_2$O system at 25 °C. *Corros. Sci.* **1973**, *13*, 659–676. [CrossRef]
40. Langova, S.; Lesko, J.; Matysek, D. Selective leaching of zinc from zinc ferrite with hydrochloric acid. *Hydrometallurgy* **2009**, *95*, 179–182. [CrossRef]
41. Nunez, C.; Vinals, J. Kinetics of leaching of zinc ferrite in aqueous hydrochloric acid solutions. *Metall. Mater. Trans. B* **1984**, *15*, 221–228. [CrossRef]
42. Sato, T.; Nakamura, T. The stability constants of the aqueous chloro complexes of divalent zinc, cadmium and mercury determined by solvent extraction with tri-n octyl phosphine oxide. *Hydrometallurgy* **1980**, *6*, 3–12. [CrossRef]
43. Winand, R. Chloride hydrometallurgy. *Hydrometallurgy* **1991**, *27*, 285–316. [CrossRef]

© 2018 by the authors. Licensee MDPI, Basel, Switzerland. This article is an open access article distributed under the terms and conditions of the Creative Commons Attribution (CC BY) license (http://creativecommons.org/licenses/by/4.0/).

Article

Electrochemical Behaviour of Chalcopyrite in Chloride Solutions

Luis Beiza [1,2], Víctor Quezada [1,3,*], Evelyn Melo [1] and Gonzalo Valenzuela [1]

1. Laboratorio de Investigación de Minerales Sulfurados, Departamento de Ingeniería Metalúrgica y Minas, Universidad Católica del Norte, Avenida Angamos 0610, 1270709 Antofagasta, Chile; lubeiza@ucn.cl (L.B.); emelo@ucn.cl (E.M.); g.valezma@gmail.com (G.V.)
2. Hydrometallurgy Research Group, Department of Chemical Engineering, University of Cape Town, South Lane, Rondebosch 7701, South Africa
3. CPCM Research Group, Department of Materials Science and Physical Chemistry, University of Barcelona, Martí i Franquès 1, 08028 Barcelona, Spain
* Correspondence: vquezada@ucn.cl; Tel.: +56-552651024

Received: 8 November 2018; Accepted: 6 January 2019; Published: 11 January 2019

Abstract: Due to the depletion of oxidized copper ores, it necessitates the need to focus on metallurgical studies regarding sulphide copper ores, such as chalcopyrite. In this research, the electrochemical behaviour of chalcopyrite has been analysed under different conditions in order to identify the parameters necessary to increase the leaching rates. This was carried out through cyclic voltammetry tests at 1 mV/s using a pure chalcopyrite macro-electrode to evaluate the effect of scan rate, temperature, and the addition of chloride, cupric, and ferrous ions. Lastly, the feasibility of using seawater for chalcopyrite dissolution was investigated. An increase in the sweep rate and temperature proved to be beneficial in obtaining highest current densities at 10 mV/s and 50 °C. Further, an increase of chloride ions enhanced the current density values. The maximum current density obtained was 0.05 A/m^2 at concentrations of 150 g/L of chloride. An increase in the concentration of cupric ions favoured the oxidation reaction of Fe (II) to Fe (III). Finally, the concentration of chloride ions present in seawater has been identified as favourable for chalcopyrite leaching.

Keywords: chalcopyrite; voltammetry; electrochemistry; seawater; chloride

---

1. Introduction

Nowadays chalcopyrite is processed mainly via concentration by flotation and subsequent pyrometallurgy. As a result of this last stage copper anodes with 99.6% purity are obtained [1] as well as slag and SO$_2$ gas. One of the main challenges of smelters in Chile is to satisfy stricter environmental regulations. Furthermore; hydrometallurgy faces the depletion of copper oxides in operating fields and the reduction of copper grades. Therefore, the main challenge is the processing of copper sulphide ores, specifically chalcopyrite (CuFeS$_2$) leaching. Chalcopyrite has been studied through various aqueous systems such as nitrate, ammonia, chloride and sulphate, with and without the presence of bacteria [2]. This would help to reduce energy and economic costs in the industry, as well as the operational projection of hydrometallurgical plants.

In addition to being the most abundant copper sulphide mineral in the earth's crust, chalcopyrite is also very refractory to conventional leaching processes [3]. Fundamental studies of leaching and electrochemistry have indicated that the dissolution of chalcopyrite in acidic environments is an electrochemical reaction, which is dependent on the solution potential between 560–620 mV (SHE) where the chalcopyrite dissolution increases [4].

Investigations by Lu et al. [5], and later Yévenes et al. [6], have indicated that chloride ions would increase this solutions potential range, however, outside this range a product layer is formed on the mineral surface which inhibits or prevents the dissolution of chalcopyrite. Additionally, intermediate copper sulphide products, such as covellite (CuS) and chalcocite (Cu_2S), would be formed under 550 mV (SHE), and these species can only be dissolved by increasing the solution potential. A Pourbaix diagram for $CuFeS_2$ was illustrated by Cordova et al. [7] which shows that the dissolution of chalcopyrite occurs through its transformation in different intermediate sulphides (Cu_5FeS_4, CuS, Cu_2S) and the dissolution of copper from chalcopyrite requires a pH lower than 4 and an oxidizing redox potential higher than 400 mV.

Different studies have indicated that a product layer will be formed on the surface of chalcopyrite above 620 mV (SHE) [4,7,8]. The presence of those products can stop or reduce the dissolution rate of chalcopyrite [7]. The controversy over the composition of this layer continues today. Different studies have found that passivation is caused by the formation of elemental sulphur [9]. Other studies indicate that they are iron precipitates, such as iron hydroxides, particularly jarosite [10,11] and even copper polysulphides or enriched copper sulphides [3,12]. Liu et al. [13] reported the formation of $Cu_2(OH)_3Cl$ as a new passive layer formed in concentrations above 0.5 mol/L NaCl however, Senanayake [14] postulated that the formation of this product occurs at pH close to 3. Despite this, it is now widely accepted that the retardation in the rate of chalcopyrite dissolution is due to the formation of a sulphide layer, rather than elemental sulphur or iron oxides which is less reactive than chalcopyrite [8].

A recent electrochemical study [15] compared three leaching systems i.e., sulphuric, nitric, and hydrochloric acid. It was indicated that the highest current densities were obtained in chloride media, and after conducting SEM/EDS analysis and Raman microspectroscopy on the electrodes they restated that both covellite and sulphur formed on the surface of chalcopyrite. In the study conducted by Nicol et al. [16], a summary of the dissolution of chalcopyrite in chloride solutions was proposed. It is indicated that at a solutions potential below 550 mV (SHE) in the presence of copper, the chalcopyrite surface is converted into one that has the characteristics of covellite. Additionally, while the solution potential increases, the chalcopyrite and the surface layer of covellite are oxidized to produce copper (I) and elemental sulphur or thiosulphite as the initial oxidation products. It is further indicated that passivation is generated by the formation of a copper polysulphide as CuS_2 when solution potential increases between 700 and 800 mV. Furthermore, at higher solution potential (over 1 V) dissolution occurs (transpasivation) and the polysulphide layer is oxidized and eliminates the passivating layer [16]. The presence of the chloride ion has been identified as a catalyst in the copper dissolution from chalcopyrite. An alternative source of chloride ions is seawater. This resource contains 20 g/L of chloride approximately. The use of seawater in Chilean copper mining takes on greater importance due to the geographical location of the deposits. The majority of copper mining in Chile is located in the Atacama Desert (the driest place in the world). As a strategic option in copper mining in Chile, seawater as the water source is capable of sustaining about 49% of the total water required for the copper industry in Chile for 2028 [17]. Thus, the use of seawater in sulphide ore leaching processes would be beneficial given the synergy provided by the presence of chloride in the system. In the present study, the electrochemical behaviour of chalcopyrite in chloride solutions was investigated. Experiments were performed at 25 and 50 °C using seawater and deionised water as solvents. A comparison has been made with the electrochemical behaviour of chalcopyrite in deionised and seawater, under the same concentration of cupric ions, ferrous ions, and sulphuric acid.

2. Materials and Methods

For this electrochemical study, the cyclic voltammetry technique was used. A potential sweep was run from the mixed potential, initially measured by open-circuit potential tests, up to a higher limit potential. The oxidation reactions occurred during this sweep, known as the anodic sweep. This was followed by a cathodic sweep which was run from the upper potential to a defined lower potential: in this sweep the reduction reactions occurred. On completion, the cycle ended at the mixed potential [18].

2.1. Instrumentation Used

Electrochemical measurements were carried out using a Voltalab PGZ100 potentiostat (Radiometer Analytical SAS, Villeurbanne, Rhone Alpes, France) with a Voltamaster VM4 software (Version F, Radiometer Analytical SAS, Villeurbanne, Rhone Alpes, France).

The electrochemical set-up consisted of a thermostatted cell and three electrodes, which is illustrated in Figure 1. The working electrode was fabricated with a pure chalcopyrite sample. XRD analysis indicated the presence of 99% of chalcopyrite (according to Figure 2). For XRD analysis, the sample was ground in an agate mortar to a size less than 45 μm and analysed in an automatic and computerized X-ray diffractometer Siemens model D5000 (Bruker, Billerica, MA, USA), with an analysis time of one hour. The ICDD (International Centre for Diffraction Data, Version PDF-2, Bruker, Billerica, MA, USA) database was used to identify the species present, and the TOPAS (Total pattern analysis software, Version 2.1, Bruker, Billerica, MA, USA) was used for quantification. The XRD equipment uses an internal corundum standard. Additionally, atomic absorption spectrometry (AAS) analysis showed 34% of copper in the sample. The reference electrode used was Hg/HgCl (calomel) which contained a saturated solution of KCl (supplied by a radiometer) and a radiometer platinum auxiliary electrode was used, which had inert characteristics. Fifty millilitres of solution were placed into the cell for each test. The solution was stirred at 60 rpm and the temperature was regulated using a thermostatted jacket. All the potentials shown are quoted with respect to the standard hydrogen electrode (SHE).

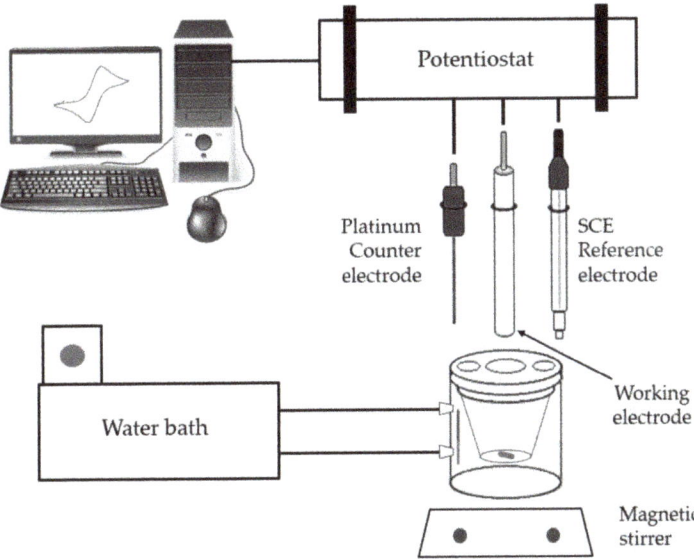

Figure 1. Experimental setup scheme.

2.2. Methodology

Open circuit potential (OCP) measurements have been carried out using both platinum and chalcopyrite electrodes in order to identify the behaviour of solution and mixed potential respectively, followed by cyclic voltammetry. The aim is to identify peak potentials in which anodic and cathodic reactions will occur.

OCP was performed for a period of 30 min before each cyclic voltammetry test in order to stabilize the rest potential. At the end of each OCP, a cyclic voltammetry test was performed with the platinum inert electrode to obtain the current densities with respect to the solution potential. This test was carried out prior to the chalcopyrite electrode tests, in order to identify the reactions that take place in the solution. The limit potentials in cyclic voltammetry tests for both the platinum inert electrode and the chalcopyrite working electrode were −500 and 1200 mV.

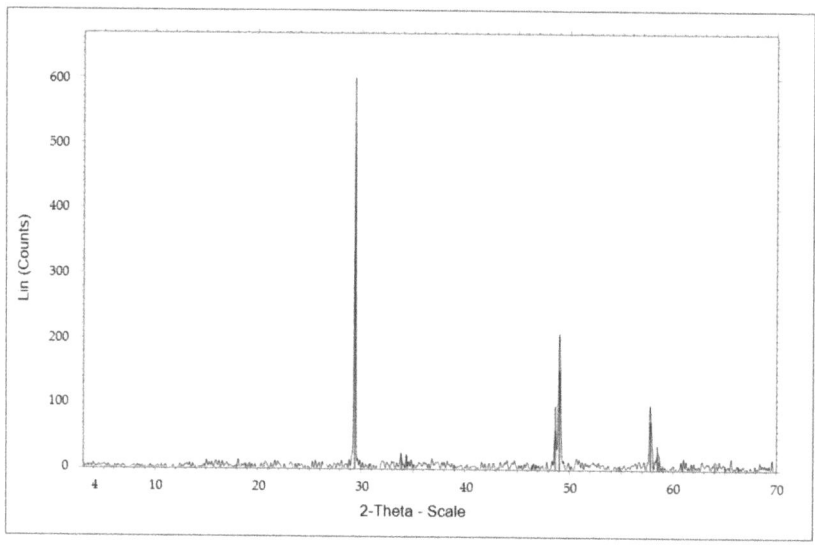

Figure 2. XRD analysis of chalcopyrite.

2.3. Parameters Assessed

The sweep rate was studied at 1, 2 and 10 mV/s using a base case solution with 20 g/L [Cl$^-$]; 0.5 g/L [Cu^{2+}]; 1 g/L [Fe^{2+}]; and pH 1 adjusted with sulphuric acid at 25 °C for all tests.

In addition, the effect of different reagents was evaluated at 25 °C using the base case solution. Chloride concentration was studied at 0, 20, 50, 100 and 150 g/L using reagent-grade NaCl. In this case, as well as the tests below, the scan rate was 1 mV/s.

The effect of ferrous ions was evaluated at 0, 1, 5, and 10 g/L [Fe^{2+}] using reagent-grade FeSO$_4$·7H$_2$O and the effect of cupric ions was studied varying the concentrations between 0; 0.5; 3 and 10 g/L [Cu^{2+}] using reagent grade CuSO$_4$·5H$_2$O.

The temperature was evaluated at 25, 35, and 50 °C using the base case electrolyte (20 g/L [Cl$^-$], 0.5 g/L [Cu^{2+}], and 1 g/L [Fe^{2+}] at pH 1).

Lastly, the effect of seawater was evaluated, using five different solutions, all adjusted to pH 1 with sulphuric acid. Seawater was collected from the seashore of Antofagasta (Chile). AAS analysis determined 19.5 g/L [Cl$^-$] in the seawater, which was filtered before the test. The electrolytes used are detailed below:

- Deionised water (DW) at pH 1
- DW at pH 1 and 20 g/L [Cl$^-$] (from NaCl)
- DW with concentrations of 20 g/L [Cl$^-$], 0.5 g/L [Cu^{2+}], 1 g/L [Fe^{2+}] at pH 1
- Seawater with concentrations of 0.5 g/L [Cu^{2+}], 1 g/L [Fe^{2+}] at pH 1

3. Results and Discussions

3.1. Effect of Sweep Speed

Figure 3 shows the anodic sweep for the chalcopyrite electrode. It can be seen that the range, where the chalcopyrite dissolution is maximized, is between 570 and 750 mV for the sweep rate of 1 and 2 mV/s. There is no significant difference between the current densities obtained, reaching a maximum of 0.05 A/m^2 at 750 mV. When 10 mV/s was used, higher current densities were obtained, reaching values close to 0.25 A/m^2 at a maximum potential of 830 mV. Under these conditions, the increase in the current density of chalcopyrite is reached within a range of 570 to 830 mV.

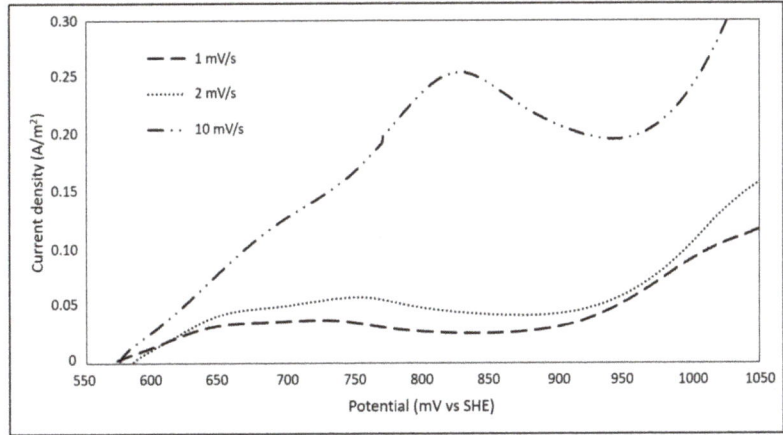

Figure 3. Effect of the sweep rate at 1, 2 and 10 mV/s on the chalcopyrite electrode using a solution with 20 g/L [Cl$^-$]; 0.5 g/L [Cu^{2+}]; 1.0 g/L [Fe^{2+}], and pH 1 adjusted with sulphuric acid at 25 °C.

According to studies conducted by Viramontes-Gamboa et al. [19], a sufficiently slow sweep speed (between 0.5 and 2 mV/s) allows the system to complete the formation of the product layer, therefore the current density will be affected by the formation of this layer on the surface of the electrode. On the other hand, using a fast enough sweep speed (over 5 mV/s), the product layer cannot be completely formed. Subsequent tests were carried out maintaining a sweep speed of 1 mV/s in order to visualize of the passivation zone, where the current density decreases within an anodic polarization.

3.2. Effect of Chloride Ions

The effect of chloride ions was analysed to compare the behaviour of current density on the chalcopyrite electrode. Figure 4 illustrates that an increase in the concentration of chloride ions up to 150 g/L [Cl$^-$] resulted an increment in current density, reaching a maximum value of 0.05 A/m^2, thus increasing the dissolution kinetics of chalcopyrite. The above is observed at solution potential of 780 mV with 150 g/L of chloride ions.

The solution potential range, where a higher current density is reached, is between 670 and 780 mV for all chloride concentrations evaluated. These results are consistent with those obtained by Lu et al. [5], who indicated that chloride ions increase the oxidation of chalcopyrite.

Similar results were obtained by Nicol et al. [16] and attributed that the oxidation peaks obtained after passivation (between 750 and 850 mV) could be associated with the oxidation of CuS$_2$ or a similar polysulphide. The above mentioned findings were obtained in the present research is presented in the Figure 5, the maximum current density obtained is over 900 mV. Similar results are obtained by Nicol [20] at potentials above 950 mV greater currents were observed in sulphate than chloride solutions.

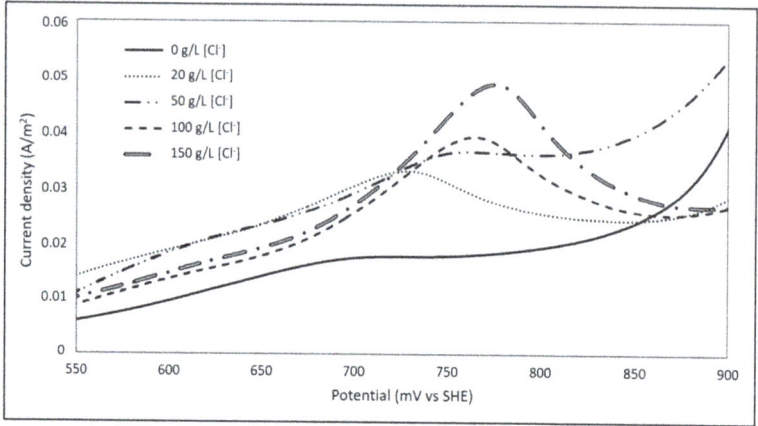

Figure 4. Effect of chloride ions on voltammograms (1 mV/s) at 0, 20, 50, 100, and 150 g/L [Cl$^-$] for the chalcopyrite electrode using a solution with 0.5 g/L [Cu^{2+}], 1.0 g/L [Fe^{2+}], and pH 1 adjusted with sulphuric acid at 25 °C.

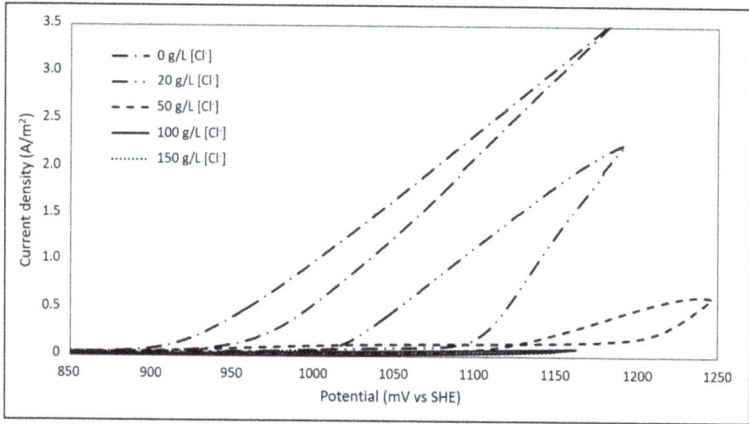

Figure 5. Effect of chloride ions in transpassivation zone (above 850 mV) on the chalcopyrite electrode (1 mV/s) at 0, 20, 50, 100, and 150 g/L [Cl$^-$] using a solution with 0.5 g/L [Cu^{2+}], 1.0 g/L [Fe^{2+}], and pH 1 adjusted with sulphuric acid at 25 °C.

3.3. Effect of Ferrous Ions

In Figure 6, two oxidation peaks are observed. The first peak corresponds to the dissolution of the chalcopyrite that occurs at approximately 670 mV, obtaining a maximum current density of 0.15 A/m^2. These results are consistent as reported by Beltrán et al. [21] under similar conditions. The above indicates that the dissolution of the chalcopyrite increases while increasing the concentration of ferrous ions in solution. A study conducted by Hiroyoshi et al. [22], observed that the chalcopyrite solution was more effective in ferrous sulphate medium than in ferric sulphate. The electrochemical studies carried out by Elsherief [3] have shown the positive effect of ferrous ions. However, other authors give more importance to the Fe (II)/Fe (III) ratio, while other studies refute this claim [6].

The second oxidation peak, which is observed at 750 mV, corresponds to the ferrous to ferric oxidation reaction. This is confirmed by the results obtained in tests with an inert electrode, which can be seen in Figure 7a.

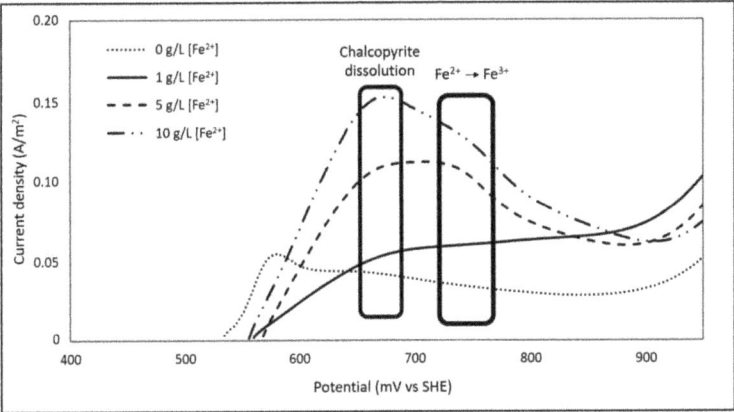

Figure 6. Effect of the concentration of ferrous ions at 0, 1, 5, and 10 g/L [Fe^{2+}] on the chalcopyrite electrode (1 mV/s) using a solution with 20 g/L [Cl^-], 0.5 g/L [Cu^{2+}], and pH 1 adjusted with sulphuric acid at 25 °C.

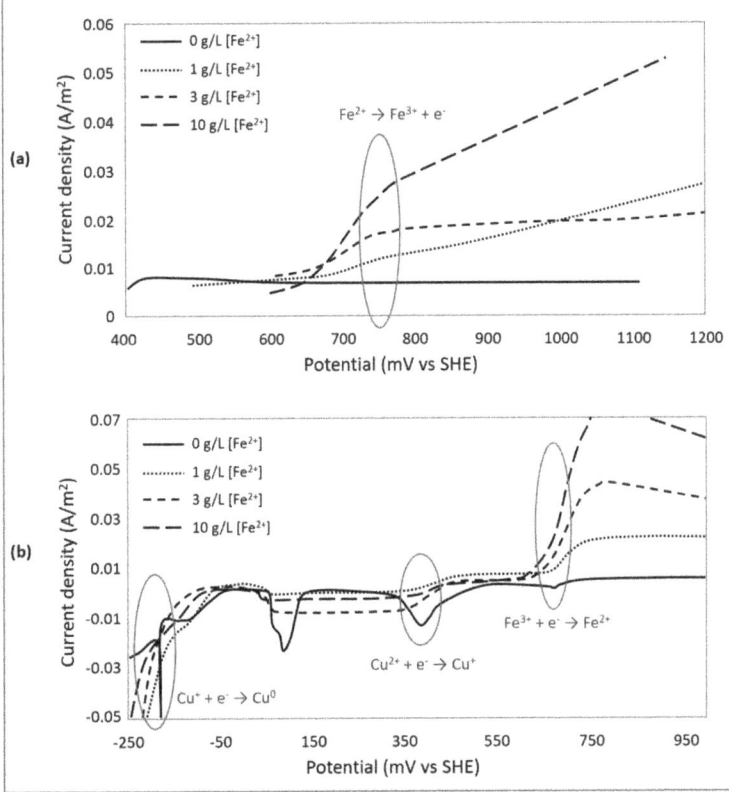

Figure 7. Effect of the concentration of ferrous ions at 0, 1, 5, and 10 g/L in the solution using an inert platinum electrode (1 mV/s) (**a**) Anodic sweep, (**b**) cathodic sweep. The solution contains 20 g/L [Cl^-], 0.5 g/L [Cu^{2+}], and pH 1 adjusted with sulphuric acid at 25 °C.

In anodic polarization, no reaction occurs in the absence of ferrous ions. On the other hand, the solution with 1 g/L of ferrous ions forms a small peak at 750 mV, and this peak is observed when the concentration of ferrous ions increases. Furthermore, Figure 7b shows three reduction peaks. The first peak at 700 mV corresponds to the reduction reaction of Fe (III) to Fe (II), the second peak at 400 mV which corresponds to the reduction reaction of Cu (II) to Cu (I) and finally at −100 mV the peak relates to the reduction reaction of Cu (I) to Cu^0. Similar results of oxidation and reduction, to the same potentials under the same solution conditions have been obtained by Beltrán et al. [21]. In addition, Hiroyoshi et al. [23], proposed a model in which an intermediate copper sulphide is formed, such as Cu_2S in the region below a critical potential, which depends on the concentration of the ferrous and cupric ions.

3.4. Effect of Cupric Ions

The results obtained are shown in Figure 8. It is observed that a rise in cupric concentration, from 0 to 10 g/L, generates an increase in current density. To perform this analysis it is necessary to identify the oxidation peaks that occur in the system. Figure 8 shows two peaks of oxidation, the first would correspond to the dissolution of chalcopyrite at potentials close to 670 mV and an oxidation potential of ferrous ions that occurs at potentials close to 750 mV.

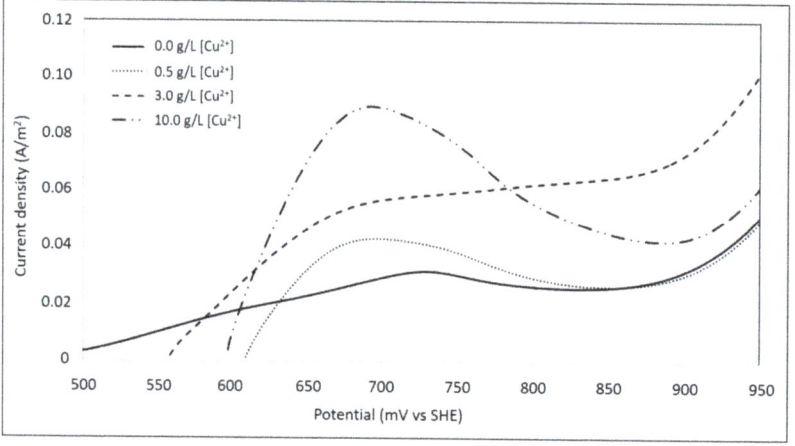

Figure 8. Effect of the concentration of cupric ions at 0, 0.5, 3.0, and 10.0 g/L on the chalcopyrite electrode (1 mV/s) using a solution with 20 g/L [Cl$^-$], 1 g/L [Fe^{2+}], and pH 1 adjusted with sulphuric acid at 25 °C.

These results are compared with an inert platinum electrode test (Figure 9a). The test determines a peak of ferrous ion oxidation near the 750 mV potential, which agrees with the findings as obtained by Beltrán et al. [21]. An increase in the concentration of cupric ion generated an increment in the current density corresponding to the oxidation peak of Fe (II) to Fe (III). This indicates that the presence of copper ions favours the oxidation of the ferrous ion acting as catalyst of the reaction.

The research of Hiroyoshi et al. [24] indicated that the coexistence of copper and ferrous ion promotes the dissolution of chalcopyrite at a potential lower than 550 mV (SHE). However, the absence of these ions, generates a passivating layer of high resistance to the dissolution of chalcopyrite. In the study by Yévenes et al. [6] tests were performed using a 0.2 M HCl solution with different concentrations of cupric ions. Results indicate that an increase in the concentration of cupric ions does not generate an increase in the rate of leaching, but a small amount (0.1 g/L) of cupric ions catalyses the leaching of copper sulphides such as chalcopyrite.

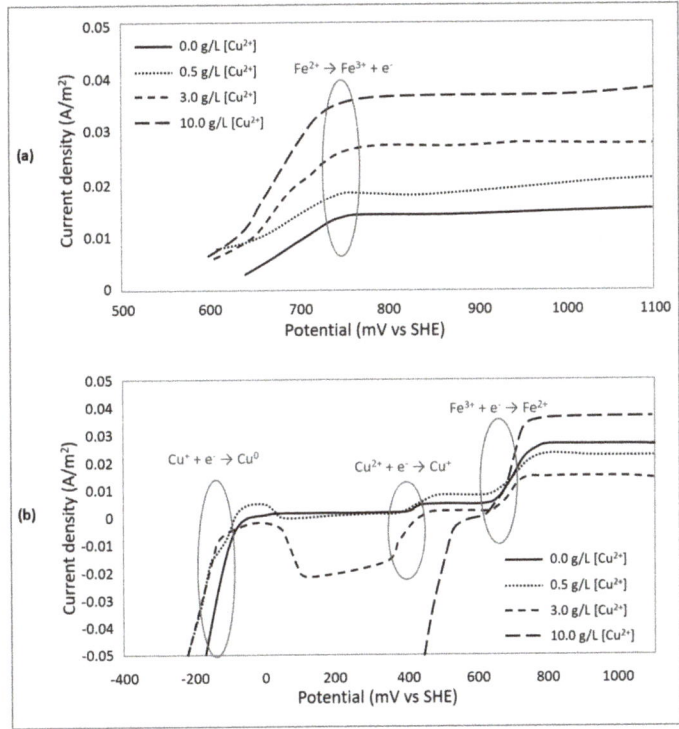

Figure 9. Effect of cupric ions at 0, 0.5, 3, and 10 g/L in the solution using an inert platinum electrode (1 mV/s) during (**a**) anodic sweep, and (**b**) cathodic sweep. The solution contains 20 g/L [Cl$^-$], 1 g/L [Fe^{2+}], and pH 1 adjusted with sulphuric acid at 25 °C.

Recent studies by Veloso et al. [25] developed leaching tests in chloride media and evaluated the effect of cupric and ferric ions. They achieved twice as much dissolution in tests with cupric ions compared tests with ferric ions. The authors indicate that this is due to faster and reversible reduction of Cu (II) to Cu (I) compared to the Fe (II)/Fe (III) couple.

In addition, the stability of Cu (I) in solution is due to its strong association to form chlor-complexes, which have greater stability with Cu (I) than iron ions [26,27]. These complexes would rapidly oxidize to Cu (II), resulting in an oxidizing agent in the reaction [6].

Figure 9b shows the curve in the cathode direction, i.e., where the reduction reactions occur. The peaks of the current density obtained indicate the potentials where the reduction reactions of Fe (III) to Fe (II) and Cu (II) to Cu (I) and then to Cu^0 occur. The peaks obtained in this curve coincide with the potentials obtained by Beltrán et al. [21].

3.5. Effect of Temperature

Figure 10 illustrates that the current density increases while the temperature increases from 25 to 50 °C, therefore, the dissolution kinetics of chalcopyrite will increment. A maximum current density of 0.17 A/m^2 is obtained at a potential of 750 mV and the range in which the chalcopyrite solution enhance is between 550 and 750 mV for all the tests performed. The trend of these results correlates with those obtained by Ibáñez and Velásquez [28] in flask leaching tests and also with those obtained by Lu et al. [5] who performed electrochemical tests at 30 and 70 °C. In the study conducted by Lu et al. [5], authors determined the activation energy through the Arrhenius equation, indicating that the chalcopyrite solution under these conditions is controlled by the chemical reaction.

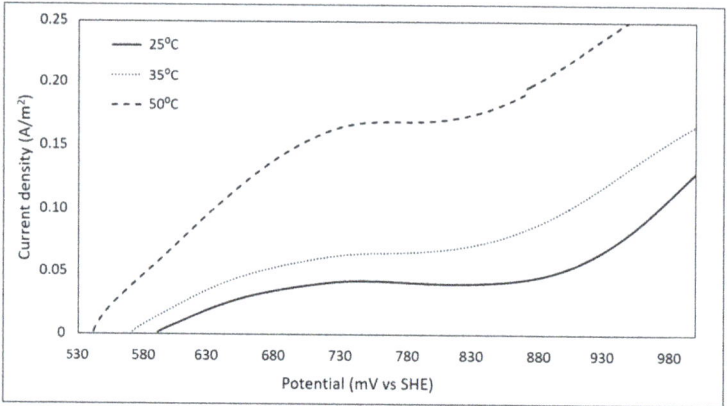

Figure 10. Effect of temperature at 25, 35, and 50 °C on the chalcopyrite electrode (1 mV/s) using a solution with 20 g/L [Cl$^-$]; 0.5 g/L [Cu^{2+}], 1.0 g/L [Fe^{2+}], and pH 1 adjusted with sulphuric acid.

3.6. Effect of Seawater

In order to evaluate the behaviour of chalcopyrite, electrochemical tests are developed using seawater (SW) and deionised water (DW). The results indicate that a maximum current density of 0.075 A/m^2 at a potential of 800 mV was obtained in the test conducted with seawater at pH 1, 0.5 g/L of Cu^{2+} and 1 g/L of Fe^{2+}. Tests performed with deionised water reached 0.020 A/m^2 for the same 800 mV. In addition, it is observed that the test performed with deionised water at pH 1, 20 g/L of Cl$^-$, 0.5 g/L of Cu^{2+}, and 1 g/L of Fe^{2+} (simulates synthetic seawater) had a similar behaviour to the seawater test under the same conditions (see Figure 11).

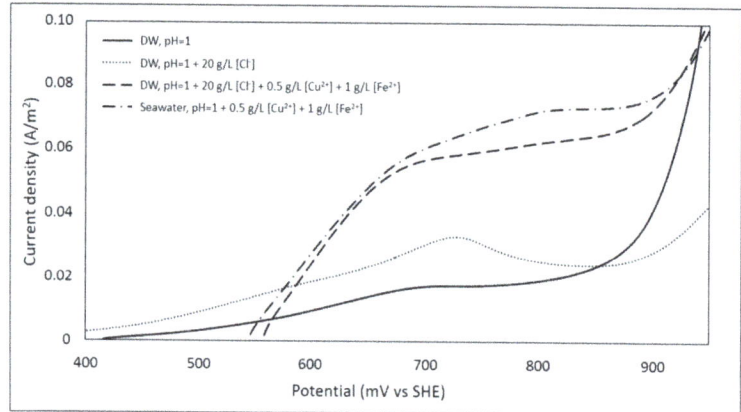

Figure 11. Effect of seawater on the chalcopyrite electrode (1 mV/s) using four different solutions at pH 1 adjusted with H$_2$SO$_4$; deionised water (DW) with no addition of any reagent; DW adding 20 g/L [Cl$^-$]; DW adding 20 g/L [Cl$^-$], 0.5 g/L [Cu^{2+}], and 1.0 g/L [Fe^{2+}]; and Seawater adding 0.5 g/L [Cu^{2+}] and 1.0 g/L [Fe^{2+}].

Test carried out with seawater, reaches a slightly higher current density than the test performed with simulated seawater. It is necessary to mention that seawater can have up to 0.2 mg/L of silver, and this ion acts as a catalyst for the dissolution of chalcopyrite [29]. However, during the present investigation the concentration of silver in solution was not measured.

These results indicate that it is possible to use seawater for chalcopyrite leaching processes. These results coincide with the study by Velásquez-Yévenes and Quezada-Reyes [30], who, in addition to the use of seawater, evaluated the use of discard brine in the chalcopyrite solution. In addition, Torres et al. [31] indicated that seawater can replace fresh water without affecting copper production.

4. Conclusions

According to results obtained, increasing the sweep rate from 1 to 10 mV/s it seems to inhibit the formation of a product layer on the surface of the chalcopyrite electrode. Therefore, this generates an increase of 0.25 A/m^2 in the current density, compared to 0.035 A/m^2 that is generated with a sweep rate of 1 mV/s.

The increase in chloride ion concentration from 0 g/L to 100 g/L in the solution increases the current density from 0.015 to 0.05 A/m^2.

In the absence of copper and iron ions, 0.03 and 0.05 A/m^2 were obtained, respectively, and in the presence of 10 g/L of these ions, an increase of 0.09 A/m^2 and 0.15 A/m^2 was observed, respectively. In addition, the cupric ions function as a catalyst for the dissolution of chalcopyrite and for the oxidation reaction of Fe (II) to Fe (III).

The concentration of chloride ion (20 g/L) that seawater possesses proved to be beneficial for the dissolution of copper from chalcopyrite. For the tests carried out, the seawater at pH 1, 0.5 g/L of Cu^{2+} and 1 g/L of Fe^{2+}, generates a current density close to 0.075 A/m^2, whereas, 0.02 A/m^2 is obtained in deionised water at pH 1 without the addition of copper and ferrous ions.

Author Contributions: Conceptualization: L.B. and V.Q.; methodology: L.B. and G.V.; validation: L.B., V.Q., and E.M.; formal analysis: L.B., V.Q., and E.M.; investigation: L.B.; resources: L.B. and V.Q.; data curation: G.V.; writing—original draft preparation: V.Q. and G.V.; writing—review and editing: L.B. and E.M.; visualization: L.B.; supervision: V.Q.; project administration: E.M.; funding acquisition: L.B. and V.Q.

Funding: This research received no external funding.

Acknowledgments: This work has been supported by the Sulphide Minerals Research Laboratory (Universidad Católica del Norte). Authors are grateful to Lilian Velásquez Yévenes (Universidad de Santiago de Chile), University of Barcelona and University of Cape Town for being part of our growth as researchers.

Conflicts of Interest: The authors declare no conflict of interest. The funders had no role in the design of the study; in the collection, analyses, or interpretation of data; in the writing of the manuscript, or in the decision to publish the results.

References

1. Cifuentes, G.; Vargas, C.; Simpson, J. Análisis de las principales variables de proceso que influyen en el rechazo de los cátodos durante el electrorrefino del cobre. *Rev. Metal.* **2009**, *45*, 228–236. [CrossRef]
2. Dreisinger, D. Copper leaching from primary sulfides: Options for biological and chemical extraction of copper. *Hydrometallurgy* **2006**, *83*, 10–20. [CrossRef]
3. Elsherief, A.E. The influence of cathodic reduction, Fe^{2+} and Cu^{2+} ions on the electrochemical dissolution of chalcopyrite in acidic solution. *Miner. Eng.* **2002**, *15*, 215–223. [CrossRef]
4. Velásquez-Yévenes, L.; Nicol, M.; Miki, H. The dissolution of chalcopyrite in chloride solutions: Part 1. the effect of solution potential. *Hydrometallurgy* **2010**, *103*, 108–113. [CrossRef]
5. Lu, Z.; Jeffrey, M.; Lawson, F. An electrochemical study of the effect of chloride ions on the dissolution of chalcopyrite in acidic solutions. *Hydrometallurgy* **2000**, *56*, 145–155. [CrossRef]
6. Yévenes, L.V.; Miki, H.; Nicol, M. The dissolution of chalcopyrite in chloride solutions: Part 2: Effect of various parameters on the rate. *Hydrometallurgy* **2010**, *103*, 80–85. [CrossRef]
7. Córdoba, E.M.; Muñoz, J.A.; Blázquez, M.L.; González, F.; Ballester, A. Leaching of chalcopyrite with ferric ion. Part I: General aspects. *Hydrometallurgy* **2008**, *93*, 81–87. [CrossRef]
8. Nicol, M.; Zhang, S. The anodic behaviour of chalcopyrite in chloride solutions: Potentiostatic measurements. *Hydrometallurgy* **2017**, *167*, 72–80. [CrossRef]
9. Carneiro, M.F.C.; Leão, V.A. The role of sodium chloride on surface properties of chalcopyrite leached with ferric sulphate. *Hydrometallurgy* **2007**, *87*, 73–82. [CrossRef]

10. Córdoba, E.M.; Muñoz, J.A.; Blázquez, M.L.; González, F.; Ballester, A. Leaching of chalcopyrite with ferric ion. Part II: Effect of redox potential. *Hydrometallurgy* **2008**, *93*, 88–96. [CrossRef]
11. Lu, Z.Y.; Jeffrey, M.I.; Lawson, F. Effect of chloride ions on the dissolution of chalcopyrite in acidic solutions. *Hydrometallurgy* **2000**, *56*, 189–202. [CrossRef]
12. Hackl, R.P.; Dreisinger, D.B.; Peters, E.; King, J.A. Passivation of chalcopyrite during oxidative leaching in sulfate media. *Hydrometallurgy* **1995**, *39*, 25–48. [CrossRef]
13. Liu, Q.; Chen, M.; Yang, Y. The effect of chloride ions on the electrochemical dissolution of chalcopyrite in sulfuric acid solutions. *Electrochim. Acta* **2017**, *253*, 257–267. [CrossRef]
14. Senanayake, G. A review of chloride assisted copper sulfide leaching by oxygenated sulfuric acid and mechanistic considerations. *Hydrometallurgy* **2009**, *98*, 21–32. [CrossRef]
15. Almeida, T.D.C.; Garcia, E.M.; Da Silva, H.W.A.; Matencio, T.; Lins, V.D.F.C. Electrochemical study of chalcopyrite dissolution in sulfuric, nitric and hydrochloric acid solutions. *Int. J. Miner. Process.* **2016**, *149*, 25–33. [CrossRef]
16. Nicol, M.; Miki, H.; Zhang, S. The anodic behaviour of chalcopyrite in chloride solutions: Voltammetry. *Hydrometallurgy* **2017**, *171*, 198–205. [CrossRef]
17. COCHILCO. *Proyección de Consumo de Agua en la Minería del Cobre 2017–2028*; COCHILCO: Santiago, Chile, 2017.
18. Baeza Reyes, A.; Mendoza, A.G. *Principios de Electroquímica Analítica*; Universidad Nacional Autónoma de México: Ciudad de México, Mexico, 2011.
19. Viramontes-Gamboa, G.; Rivera-Vasquez, B.F.; Dixon, D.G. The Active-Passive Behavior of Chalcopyrite. *J. Electrochem. Soc.* **2007**, *154*, C299. [CrossRef]
20. Nicol, M.J. The anodic behaviour of chalcopyrite in chloride solutions: Overall features and comparison with sulfate solutions. *Hydrometallurgy* **2017**, *169*, 321–329. [CrossRef]
21. Beltrán, E.C.; Frisch, G.; Velasquez, L. Chalcopyrite Dissolution using Electrochemical Tests. In Proceedings of the Sustainable Hydrometallurgical Extraction of Metals, SAIMM, Cape Town, South Africa, 1–3 August 2016; pp. 157–165.
22. Hiroyoshi, N.; Miki, H.; Hirajima, T.; Tsunekawa, M. Enhancement of chalcopyrite leaching by ferrous ions in acidic ferric sulfate solutions. *Hydrometallurgy* **2001**, *60*, 185–197. [CrossRef]
23. Hiroyoshi, N.; Miki, H.; Hirajima, T.; Tsunekawa, M. Model for ferrous-promoted chalcopyrite leaching. *Hydrometallurgy* **2000**, *57*, 31–38. [CrossRef]
24. Hiroyoshi, N.; Kuroiwa, S.; Miki, H.; Tsunekawa, M.; Hirajima, T. Synergistic effect of cupric and ferrous ions on active-passive behavior in anodic dissolution of chalcopyrite in sulfuric acid solutions. *Hydrometallurgy* **2004**, *74*, 103–116. [CrossRef]
25. Veloso, T.C.; Peixoto, J.J.M.; Pereira, M.S.; Leao, V.A. Kinetics of chalcopyrite leaching in either ferric sulphate or cupric sulphate media in the presence of NaCl. *Int. J. Miner. Process.* **2016**, *148*, 147–154. [CrossRef]
26. Senanayake, G. Chloride assisted leaching of chalcocite by oxygenated sulphuric acid via Cu(II)-OH-Cl. *Miner. Eng.* **2007**, *20*, 1075–1088. [CrossRef]
27. Velásquez Yévenes, L. The Kinetics of the Dissolution of Chalcopyrite in Chloride Media. Ph.D. Thesis, Murdoch University, Perth, Australia, 2009.
28. Ibáñez, T.; Velásquez, L. Lixiviación de la calcopirita en medios clorurados. *Rev. Metal.* **2013**, *49*, 131–144. [CrossRef]
29. Zierenberg, R.A.; Schiffmant, P. Microbial control of silver mineralization at a sea-floor hydrothermal site on the northern Gorda Ridge. *Nature* **1990**, *348*, 155–157. [CrossRef]
30. Velásquez-Yévenes, L.; Quezada-Reyes, V. Influence of seawater and discard brine on the dissolution of copper ore and copper concentrate. *Hydrometallurgy* **2018**, *180*, 88–95. [CrossRef]
31. Torres, C.M.; Taboada, M.E.; Graber, T.A.; Herreros, O.O.; Ghorbani, Y.; Watling, H.R. The effect of seawater based media on copper dissolution from low-grade copper ore. *Miner. Eng.* **2015**, *71*, 139–145. [CrossRef]

© 2019 by the authors. Licensee MDPI, Basel, Switzerland. This article is an open access article distributed under the terms and conditions of the Creative Commons Attribution (CC BY) license (http://creativecommons.org/licenses/by/4.0/).

Article

A Mineralogical Assessment on Residues after Acidic Leaching of Bauxite Residue (Red Mud) for Titanium Recovery

Gözde Alkan [1,*], Claudia Schier [1], Lars Gronen [2], Srecko Stopic [1] and Bernd Friedrich [1]

1. IME-Process Metallurgy and Metal Recycling, RWTH Aachen University, Intzestraße 3, 52056 Aachen, Germany; cschier@ime-aachen.de (C.S.); sstopic@ime-aachen.de (S.S.); bfriedrich@ime-aachen.de (B.F.)
2. IML-Chair of Applied Mineralogy and Economic Geology, RWTH Aachen University, Wüllnerstraße 2, 52062 Aachen, Germany; gronen@emr.rwth-aachen.de
* Correspondence: galkan@ime-aachen.de; Tel.: +49-24195873

Received: 27 September 2017; Accepted: 23 October 2017; Published: 28 October 2017

Abstract: Due to its alkalinity, red mud produced by the Bayer process may affect both the environment and human health. For this reason, its further utilization instead of disposal is of great importance. Numerous methods have already been studied for hydrometallurgical treatment of red mud, especially for the recovery of various metallic components such as iron, aluminum, titanium or rare earth elements. This study focuses on the extraction of titanium from red mud and in particular the mineralogical changes, induced by leaching. Sulfuric acid, hydrochloric acid and their combination have been utilized as leaching agents with the same leaching parameters. It has been determined that sulfuric acid is the best candidate for the red mud treatment in terms of titanium leaching efficiency at the end of 2 h with a value of 67.3%. Moreover, samples from intermediate times of reaction revealed that leaching of Ti exhibit various reaction rates at different times of reaction depending on acid type. In order to explain differences, X-ray Diffraction (XRD), scanning electron microscope (SEM) and QEMSCAN techniques were utilized. Beside titanium oxide (TiO_2) with available free surface area, a certain amount of the TiO_2 was detected as entrapped in Fe dominating oxide. These associations between Ti and Fe phases were used to explain different leaching reaction rates and a reaction mechanism was proposed to open a process window.

Keywords: bauxite residue; red mud; leaching; titanium; metal recovery

---

1. Introduction

There are growing efforts in industry to promote the sustainability and implementation of zero-waste production; the use of waste products from industrial processes is becoming increasingly important. During alumina production by Bayer process, a large amount of bauxite residue (red mud) is formed as a waste product [1–3]. The cumulative amount of red mud by 2015 is estimated to be close to 4×10^9 tons. These higher production rates and precious mineral content such as Fe_2O_3, Al_2O_3, SiO_2, TiO_2, Na_2O and CaO and rare earth elements induced the utilization of red mud as a secondary resource [4,5]. Reduced process costs with respect to primary metal production routes favors red mud usage in economic aspects. Moreover, valorization of such a highly alkaline product stored in the environment also provides ecological benefits [4]. Red mud can also be considered a secondary source of the most important modification of titanium compound, titanium dioxide. Owing to its outstanding properties, in particular its high refractive index, titanium dioxide is commonly used as white pigment in numerous fields of industry such as, dyes, plastics or even eatables and drugs [6,7]. Due to decreasing availabilities as well as qualities of the titanium ores, recovery of titanium from red mud by hydrometallurgical methods gains importance. Depending on the source, red mud may exhibit higher amount of Ti (up to 25%), which affects leaching efficiency and selectivity [6]. There have been many studies conducted on titanium recovery from red mud by hydrometallurgical methods [6]. Among other inorganic acids, sulfuric acid (H_2SO_4) is reported as the best choice for higher titanium leaching efficiencies followed by chloric acid (HCl) [4]. Ti extraction rates reached 71% in the case of sulfuric acid leaching [7].

However, there is not a systematic study to explain the differences in leaching mechanisms and kinetics depending on acid type. This lack of knowledge leads to the detailed phase and mineralogical investigation on red mud and leach residue to open a process window. Hydrochloric acid, sulfuric acid and their combination were utilized as leachate with identical leaching conditions. Differences and similarities in leaching efficiencies and kinetics revealed by inductively plasma optical emission spectometryanalysis were explained using X-ray Diffraction (XRD), scanning electron microscope (SEM) and QEMSCAN techniques. A new leaching model was proposed for HCl and H_2SO_4 and difficulties of titanium dissolution from red mud were explained in detail.

2. Materials and Methods

2.1. Chemical Composition of Red Mud

The red mud used for this study was generated in Aluminum of Greece (Boeotia, Greece) S.A. The chemical composition of the red mud is listed in Table 1 and shows the major constituents of this material such as iron (III)-oxide, aluminum oxide, calcium oxide and silicon dioxide analyzed by inductively plasma Optical Emission Spectrometry.

Table 1. Chemical composition of red mud.

Composition	wt %
Fe_2O_3	42.34
Al_2O_3	16.26
Ignition loss	12.66
CaO	11.64
SiO_2	6.97
TiO_2	4.27
Na_2O	3.83
Others	1.85
La_2O_3	0.09
CeO_2	0.06
Sc_2O_3	0.02
Nd_2O_3	0.01
Y_2O_3	0.01

2.2. Experimental Method

A typical leaching test was performed with a heating plate and magnetic stirrer to control the temperature and stirring speed. The leaching efficiencies of sulfuric acid, hydrochloric acid and a combination of the two acids in the ratio of hydrochloric acid to sulfuric acid of 1:3 were investigated with the same leaching parameters. The experiments were carried out at a set temperature of 70 °C, 360 rpm and a 4-molar acid concentration. The solid-liquid ratio was adjusted to a ratio of 1:50 in order to prevent silica gel formation. In order to reveal leaching rates as a function of reaction time, for all three acids, leaching duration was varied as 5, 15, 30, 60, 90 and 120 min. At the end of reaction time, samples were vacuum filtrated for solid and liquid separation. Solid residues were dried overnight at 90 °C and prepared for characterization. For XRD analyses, dried particles were milled to prevent scattering effect. However, they were analyzed as dried version by SEM and QEMSCAN so as not to retain available free surface areas of minerals. Leach liquor was diluted 20 times for ICP analyses.

2.3. Materials Characterization

Automated quantitative mineralogy analyses of red mud and solid residues were performed using a Qanta-650 F (FEI corporate headquarters, North America Nanoport, Hillsboro, OR, USA) QEMSCAN scanning electron microscope (SEM). The particular samples were embedded into epoxy resin to form blocks of approx. 25.4 mm in diameter. For analysis by SEM, the sample surface was polished and carbon coated to give best analytical results and to avoid surface charging during electron bombardment [8]. The system mounts 2 DualX-Flash (Bruker AXS, Karlsruhe, Germany) energy-disperse detectors for recording X-ray spectra emitted by the interaction of the electron beam with the atoms of the sample surface. Additionally, a 4 quadrant backscatter electron and a secondary electron (SE) detector are installed for image acquisition.

For QEMSCAN analysis, the acceleration voltage was set to 25 kV. The sample current was set to 10 nA at a working distance of 13 mm. The point spacing was set to 7.5 µm per step and 2000 X-ray counts were recorded per step. Phase interpretation and further image analysis, like phase map, modal composition and elemental mapping, were performed by using iDiscover software suite (FEI).

Back scattered electron (BSE) and secondary electron (SE) image acquisition were performed by the same system in SEM-mode using an acceleration voltage of 15 kV. For a better resolution of the images the working distance was decreased to 11.7 µm.

XRD analyses was performed by Bruker D8 Advanced Diffractometer (Bruker AXS, Karlsruhe, Germany), which use Bragg-Brentano Geometry and θ–θ synchronization for X-Ray tube and the detector. 10°–80° (2θ) were scanned with a 5°/min rate. The generator voltage was 40 kV and current was 40 nA. The quantitative evaluation was carried out with the program Topas (Bruker, AXS, Karlsruhe, Germany). For Rietveld analysis, the full profile method was used; 0-point errors, specimen height errors, diffraction-peak intensities, Lorantz Polarisation (LP)-factor, background and crystal structure models were refined.

3. Results and Discussion

The evaluation of titanium leaching efficiencies as a function of reaction time can be seen in Figure 1. The graph represents the average leaching values, including deviations due to test repetitions. At the end of two hours, the highest leaching efficiency of titanium (67.3%) was achieved with sulfuric acid, followed by hydrochloric acid (59.8%) and the combination of HCl to H_2SO_4 with a ratio of 1:3 (56.4%).

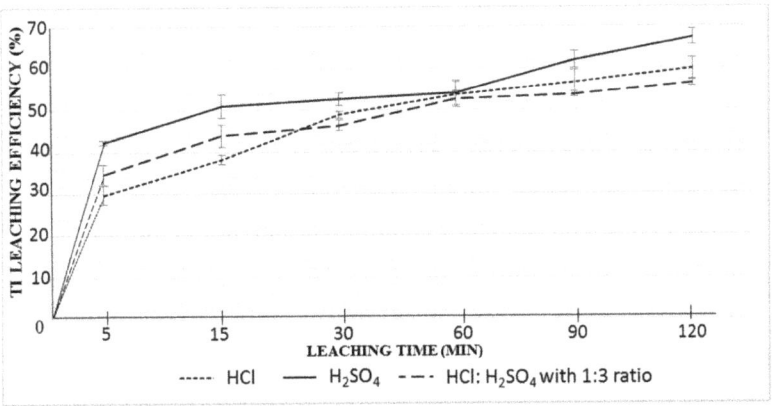

Figure 1. Leaching efficiencies of Ti with various acids (where 100% Ti efficiency corresponds to 0.94 g Ti in 1 L acid solution).

Moreover, the time dependent evaluation of leaching efficiencies revealed variable kinetic profiles after three different acid treatments. In the first 15 min of leaching, for all acid types, higher leaching rates were observed by a higher slope of the curves. After 15 min, in the case of H_2SO_4, the leaching rate decreases until 60 min test duration. In contrast, during this period (15–60 min), there is still high leaching rates observed with hydrochloric acid. After 60 min, it is seen that the HCl leaching curve gets closer to saturation and leaching becomes slower, while H_2SO_4 reaches higher rates and superior leaching efficiency. The course of the combination of hydrochloric- and sulfuric acid is characterized by less significant leaps. From this data it is seen that the most efficient leachate in terms of final leaching efficiency for Ti is sulfuric acid. However, the findings on leaching kinetics induced the detailed investigation of leaching to explain different mechanisms by different leachates and find an optimum process condition.

Using X-ray diffraction, the phase content of solid residues after HCl and H_2SO_4 treatment were investigated to reveal the effect of acid type on leaching mechanism. Related XRD analyses are given in Figure 2a,b for direct comparison purposes.

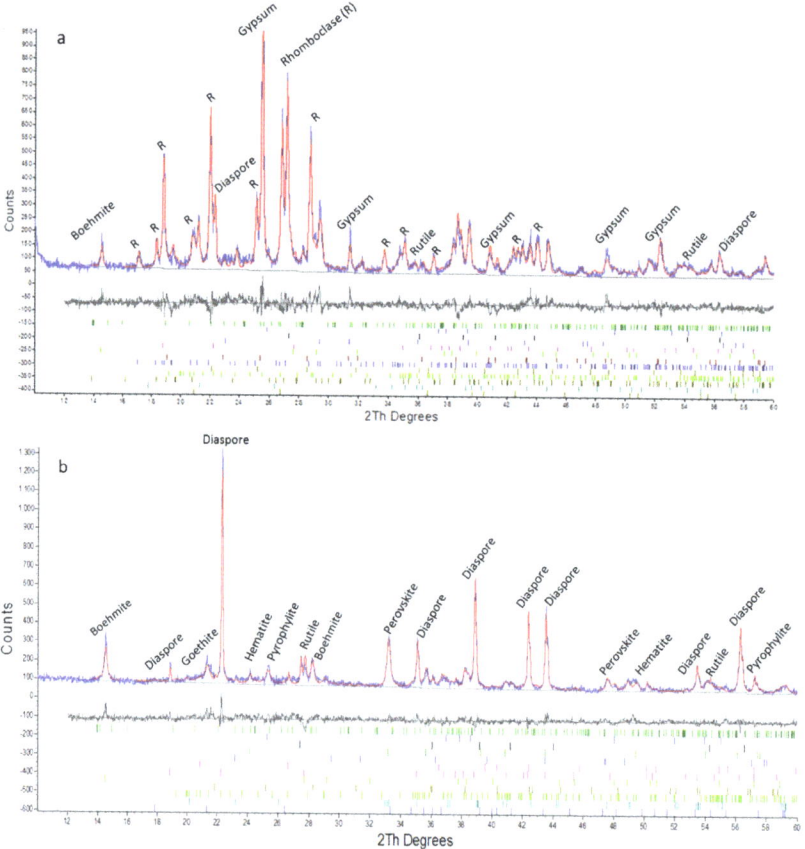

Figure 2. X-ray Diffraction (XRD) analyses of solid residues at the end of 2 h: (**a**) H_2SO_4 and (**b**) HCl leaching with Rietveld refinement (blue line is experimental, red line is calculated and the difference below with marked reflection positions).

Table 2. Quantitative analyses of sulfuric acid (H_2SO_4) and chloric acid (HCl) residues.

Composition	H_2SO_4 Residue (%)	HCl Residue (%)
Rutile (TiO_2)	0.58	1.50
Anatase (TiO_2)	0.81	0
Perovskite ($CATiO_3$)	0	3
Hematite (Fe_2O_3)	0	5.3
Goethite (FeO(OH))	0.167	6.3
Rhomboclase ($HFe(SO_4)_2 \cdot 4(H_2O)$)	57.4	0
$FeSO_4 \cdot 6H_2O$ *	3.47	0
Diaspore (AlO(OH))	15.39	60
Boehmite (AlO(OH))	2.75	10
Pyrophylite ($Al_2[(OH)_2] | Si_4O_{10}$)	1.8	9.4
Quartz (SiO_2)	0	0.1
Crotobalite (SiO_2)	0.61	0.193
Gypsum ($CaSO_4 \cdot 2H_2O$)	12.45	0
Albite ($Na(AlSi_3O_8)$)	2.4	3.06

* unnamed, possibly an Fe-analogue of retgersite.

XRD analyses revealed the differences in phase contents of leach residues in terms of titanium (Ti), iron (Fe) including phases, which may explain the different leaching behavior of two acids. As listed in Table 2, total amount of Fe including phases (hematite (5.3%) and goethite (6.3%)) are low in HCl solid residue. Nevertheless, it is seen that H_2SO_4 residue is much more enriched in terms of Fe compounds such as, rhomboclase, another iron sulfate and trace amounts of goethite. A higher amounts of Fe is found as rhomboclase (57.4%) which is formed by re-precipitation of dissolved Fe ions with sulfate. In the presence of sulfate ions and highly acidic conditions provided by concentrated H_2SO_4 leachate, the precipitation of dissolved Fe into rhomboclase is thermodynamically more favorable and may result in less leaching efficiency of Fe [9]. Quantitative analyses of residues imply in parallel with ICP analyses, higher Fe dissolution rates in the presence of HCl. Moreover, at the end of 2 h HCl leaching, Ti was detected in HCl leach residue in the form of perovskite (3%) and rutile (1.5%). In contrast, Ti was found only as rutile (0.58%) and anatase (0.81%) in the H_2SO_4 leach residue with relatively lower amounts (see Table 2), implying higher leaching efficiencies in parallel with ICP findings. Perovskite may be consumed by the reaction of calcium with sulfate ions and precipitate into gypsum as revealed in X-ray diffractogram of H_2SO_4 slag; which may also favor Ti extraction from perovskite. Higher Ti and lower Fe leaching efficiencies with H_2SO_4 favor its utilization for selective Ti leaching from red mud.

Nonetheless, when 4 wt % content of Ti in red mud and even lower value for residue is considered, XRD is not highly sensitive to reveal changes in Ti including phases. This lack of sensitivity induced QEMSCAN utilization which can deal with elemental and phase mappings, phase associations and available surface area detections for red mud and the most promising leachate (H_2SO_4) residue.

In Figure 3, the Ti elemental mapping of red mud and leach residue are given together for direct comparison purposes. The field scan image of red mud in Figure 3 revealed that Ti content of red mud phases is inhomogeneous with a maximum amount of 8–9 wt %. After leaching, a dramatic change in distribution is observed, which exhibits homogenous and finely distributed Ti through the minerals of H_2SO_4 residue with a decreased amount at a maximum 3–4 wt %.

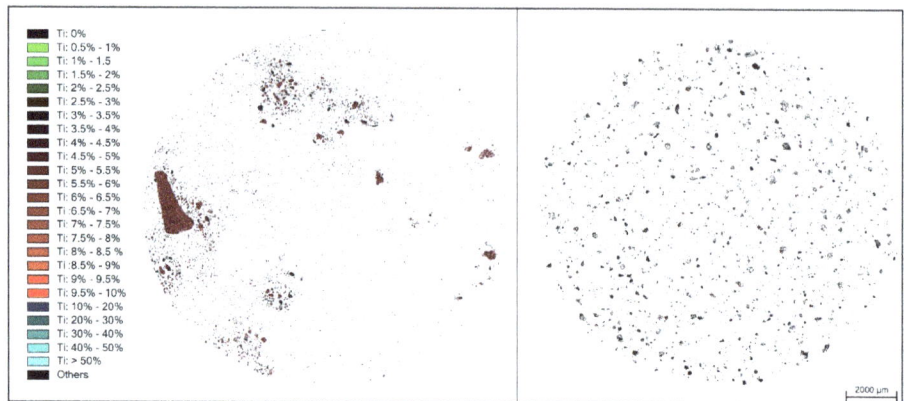

Figure 3. Ti elemental mapping of red mud (left) and H$_2$SO$_4$ leach residue (right) by QEMSCAN analyses.

Beyond elemental mapping, the mineral distribution within red mud and H$_2$SO$_4$ leach residue were also investigated in a comparative manner.

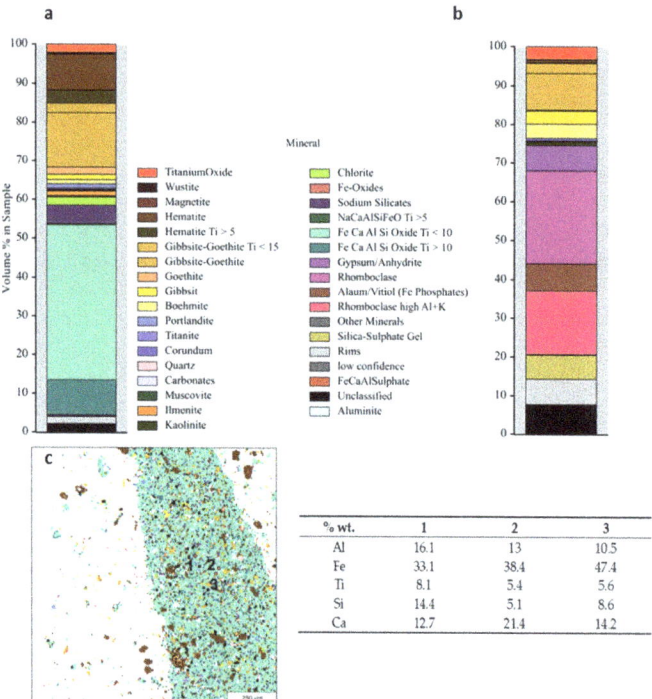

% wt.	1	2	3
Al	16.1	13	10.5
Fe	33.1	38.4	47.4
Ti	8.1	5.4	5.6
Si	14.4	5.1	8.6
Ca	12.7	21.4	14.2

Figure 4. Mineral distribution of red mud: (**a**) and H$_2$SO$_4$ leach residue (**b**) with detailed compositional analyses (**c**) reveled by QEMSCAN and table represents chemical compositions of points labeled as 1, 2 and 3.

In addition to phases indicated by XRD, QEMSCAN, analysis given in Figure 4 revealed the presence of large amounts of Fe-, Ca-, Al-, Si mixed oxide in red mud, where a certain amount of TiO$_2$ is entrapped. Due to the heterogeneous nature of this complex oxide, chemical composition and stoichiometry vary through the volume. Therefore, a crystalline phase could not be assigned. Varying compositions revealed by point 1, 2 and 3 imply that this complex oxide may be aggregate or intergrowth of several oxides inherent from Bayer Process. After leaching with H$_2$SO$_4$, as represented in the mineral distribution in Figure 4b, it is seen that most of this various composed oxide is leached out and rhomboclase formation takes place as consistent with XRD analysis. In leach residue, a limited amount of TiO$_2$ is detected only within gibbsite-goethite as revealed in Figure 4b.

The mineral association analysis revealing free available surfaces and contacts between phases is represented in Figure 5a,b for red mud and H$_2$SO$_4$ leach residue respectively.

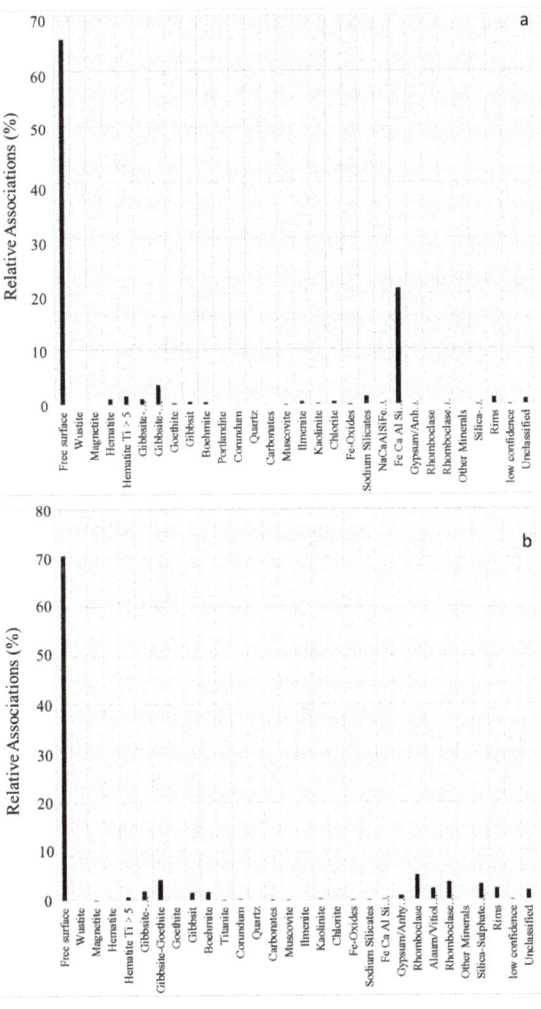

Figure 5. Mineral association of TiO$_2$ within: (a) red mud; (b) H$_2$SO$_4$ leach residue.

Figure 5a exhibits that certain amounts of TiO_2 surfaces are in contact with Fe-, Ca-, Al-, Si-oxide. The remaining TiO_2 was found with free available surfaces, which may result in fast leaching kinetics in the very early stage of leaching revealed by ICP analyses given in Figure 1. In comparison with red mud, when H_2SO_4 leach residue is considered as represented in Figure 5b, it is seen that most of the TiO_2 surfaces are free and available, just very negligible quantities are in contact with rhomboclase and gibbsite-goethite. These findings reveal that the Fe dominating complex composed oxides form a diffusion barrier between TiO_2 and leachate and may be the reason for the deceleration of reaction kinetics in the middle period of the trial.

Red mud and H_2SO_4 leach residue were investigated in a comparative manner by SEM to reveal morphological changes taking place during leaching, as represented in Figure 6. In order to reveal elemental composition, Energy dispersive X-Ray analyses (EDX) was also utilized.

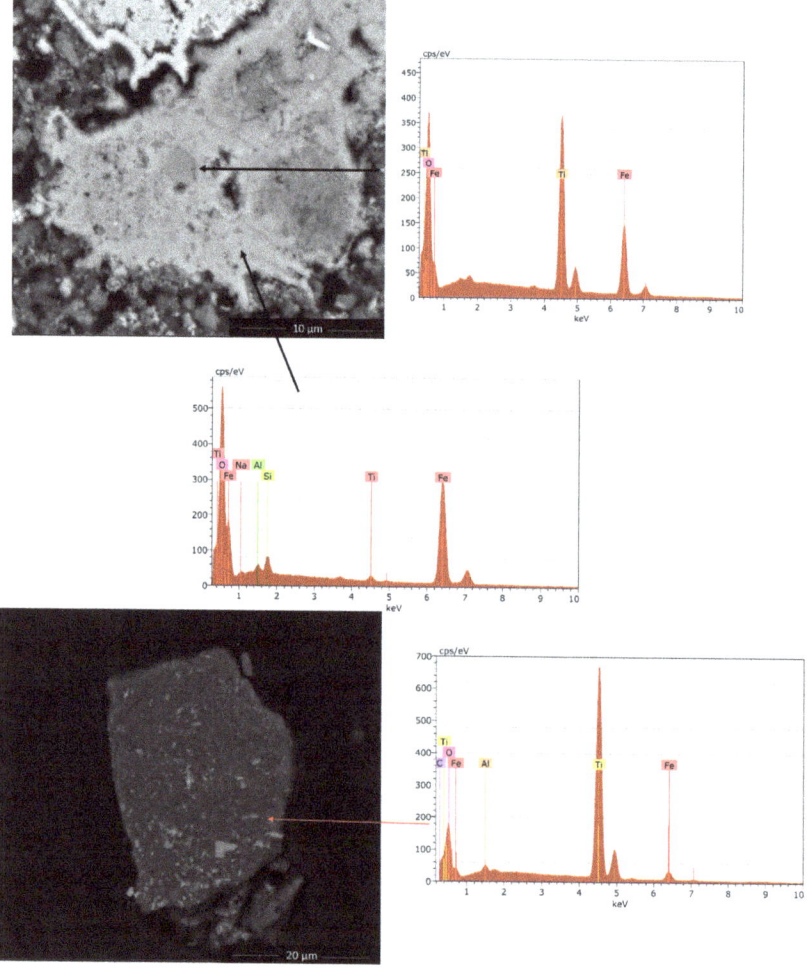

Figure 6. Scanning electron microscope (SEM) micrographs of red mud (above) and H_2SO_4 leach residue (below).

Relatively large aggregates around 30 μm of Fe (also Al, Si and Na including) dominating oxides have been revealed where 2–3 μm sized TiO$_2$ is entrapped in parallel with findings of QEMSCAN. This micrograph emphasizes that in order to reach to TiO$_2$ enriched compound, the more iron oxide enriched phase should be leached out. On the other hand, leach residue exhibited relatively finer particles with respect to coarser aggregates of red mud. Moreover, in leach residue, mostly TiO$_2$ is detected with free surfaces and small contact with Al dominating oxide where some Fe is also present, as represented in Figure 6. This is consistent with QEMSCAN analyses which indicates mostly free surfaces of TiO$_2$ and a trace amount of contact with gibbsite-goethite.

Presence of both free and entrapped TiO$_2$ surfaces revealed by SEM and QEMSCAN may be responsible for various leaching kinetic regimes within various stages of reaction. Since leaching starts from surface, initial high leaching kinetic for all acid treatments is owing to easy access to TiO$_2$ surfaces. The second regime, as observed in Figure 1, starts from 15 min of reaction where H$_2$SO$_4$ has slower and HCl similar behavior with respect to the first regime. It is worth emphasizing that as reported in previous studies, H$_2$SO$_4$ is more sensitive to TiO$_2$ leaching where HCl was used to dissolve Fe [10]. Their selectivity to metal types may be the reason for different kinetics in the second regime. Faster Fe leaching in the earlier stages of reaction by HCl ensures leachate to access TiO$_2$ entrapped in Fe dominating oxide. Relatively lower leaching rates of Fe in the case of H$_2$SO$_4$ may result in highly pronounced diffusion barrier effect of Fe enriched oxide for leaching of TiO$_2$ However, after this time H$_2$SO$_4$ achieves higher kinetics, most probably due to increased Fe leaching with longer reaction times. In the end, H$_2$SO$_4$ yields in higher Ti leaching efficiency as consistent with previous studies [10,11].

Iron oxide enriched phase inhibiting effect implied by SEM and QEMSCAN is analyzed by ICP measurement to reveal Fe leaching rates with HCl and H$_2$SO$_4$. Figure 7a,b represent leaching rates of Fe and Ti for H$_2$SO$_4$ and HCl respectively.

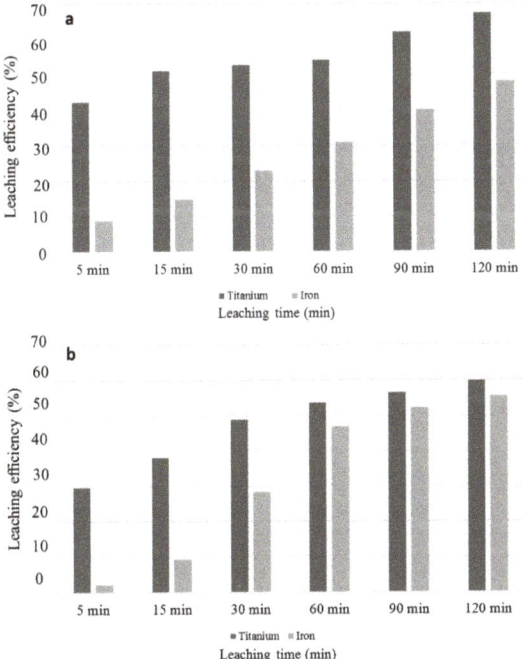

Figure 7. Titanium and iron leaching efficiencies by (**a**) sulfuric acid; (**b**) hydrochloric acid leaching (where 100% Ti and Fe leaching efficiencies corresponds to 0.94 g Ti and 8.47 g Fe in 1 L acid solution).

Figure 7a,b reveal that Fe leaching rates are similar in the early stages of leaching. However, after 15 min of reaction, Fe oxide dissolution is increased dramatically with HCl. Although they are quite similar in the early stages of reaction, after 15 min, an increase in HCl leach rate is observed while that in H_2SO_4 is still slow. After 60 min of reaction, only 30% of iron is leached out by H_2SO_4, while 46% by HCl. After 60 min, Fe leaching rate by H_2SO_4 becomes also higher and achieved to 47% at the end of 2 h, while HCl reached 55%. After 60 min, leaching rates of Ti by H_2SO_4 accelerates. Increasing Ti leaching rates with increasing Fe dissolution, in parallel with SEM and QEMSCAN analyses, indicate the obstacle effect of Fe oxide enriched mineral over Ti recovery. More leached out Fe enriched oxide favors the access of leachate to the entrapped TiO_2, which results in higher kinetics of H_2SO_4 in the last stage of the leaching.

In light of SEM, QEMSCAN and ICP analyses, a model for leaching process with HCl and H_2SO_4 are proposed as in Figure 8. This model is based on already existing particle dissolution model in leaching, where the reaction starts from surface and propogates through the core.

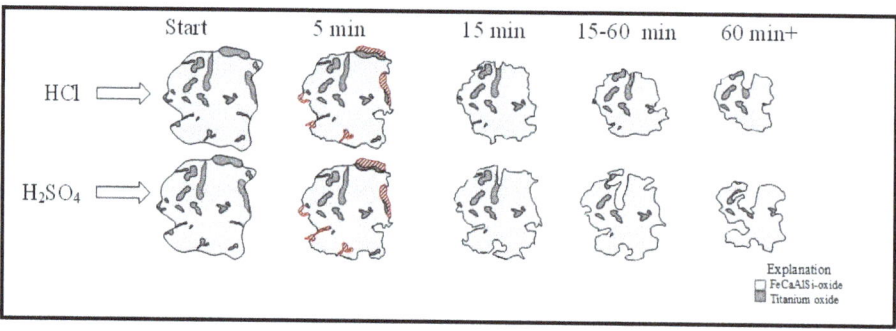

Figure 8. Dissolution model of particles by hydrochloric acid and sulfuric acid leaching.

After a period of 5 min, more TiO_2 is dissolved in sulfuric acid compared with hydrochloric acid. This can be seen in the red hatching in Figure 8. Due to the higher sensitivity of hydrochloric acid leaching to iron, in the middle time of leaching after free TiO_2 surfaces are consumed, HCl achieves entrapped TiO_2 more easily and exhibits faster kinetics. However, after 60 min, H_2SO_4 also leaches a certain amount of Fe for the exposure of TiO_2 from the Fe Ca Al Si-oxide. The remaining TiO_2 particles are leached out after a time of 60 min and at the end of reaction less amount of TiO_2 remains in residue of H_2SO_4 with respect to that of HCl.

4. Conclusions

A detailed examination of the red mud leaching was carried out to reveal mineralogical changes with various acid types in order to explain the dissolution of titanium from red mud. At the end of two hours of leaching, the highest sensitivity to titanium dissolution is observed with a value of 67% in the case of sulfuric acid leaching. Moreover, Ti and Fe leaching rates as a function of reaction time have been investigated. A mineral association between Ti and Fe and the importance of available surfaces of titanium including particles on leaching efficiency were reported. It was pointed out that Fe has an obstacle effect on Ti leaching and there should be a certain amount of Fe dissolution for better Ti leaching.

Acknowledgments: The authors wish to thank the Aachen Know-How Centre Resource Technology (AKR) for the support of an open access. This research is founded by European Community's Horizon 2020 Programme ([H2020-MSCA-ITN-2014]) under Grant Agreement No. 636876 (MSCA-ETN REDMUD). Authors also acknowledge Yiannis Pontikes, Koen Binnemans, Ken Evans and George Blagovyi for their valuable support.

Author Contributions: Claudia Schier and Gözde Alkan designed and performed the experiments; Gözde Alkan, Claudia Schier, Bernd Friedrich analyzed the data; Srecko Stopic and Lars Gronen contributed reagents/materials/analysis tools; Gözde Alkanand, Claudia Schier wrote the paper.

Conflicts of Interest: The authors declare no conflicts of interest.

References

1. Klauber, C.; Gräfe, M.; Power, G. Bauxite residue issues: II. Options for residue utilization. *Hydrometallurgy* **2011**, *108*, 11–32. [CrossRef]
2. Piga, L.; Pochetti, F.; Stoppa, L. Recovering metals from red mud generated during alumina production. *J. Miner. Met. Mater. Soc.* **1993**, *45*, 54–59. [CrossRef]
3. Zhang, W.; Zhu, Z.; Cheng, C.Y. A literature review of titanium metallurgical processes. *Hydrometallurgy* **2011**, *108*, 177–188. [CrossRef]
4. Lim, K.; Shon, B. Metal Components (Fe, Al and Ti) recovery from red mud by sulfuric acid leaching assisted with ultrasonic waves. *Int. J. Emerg. Technol. Adv. Eng.* **2015**, *5*, 25–32.
5. Agatzini-Leonardou, S.; Oustadakis, P. Titanium leaching from red mud by diluted sulfuric acid at atmospheric pressure. *J. Hazard. Mater.* **2008**, *157*, 579–586. [CrossRef] [PubMed]
6. Kasliwal, P.; Sai, P.S.T. Enrichment of titanium dioxide in red mud: kinetic study. *Hydrometallurgy* **1999**, *53*, 73–87. [CrossRef]
7. Ghorbani, A.; Fakhariyan, A. Recovery of Al_2O_3, Fe_2O_3 and TiO_2 from bauxite processing waste (red mud) by using combination of different acid. *J. Basic. Appl. Sci. Res.* **2013**, *3*, 187–191.
8. Reed, S.J.B. *Electron Microprobe Analysis and Scanning Electron Microscopy in Geology*; Cambridge University Press: Cambridge, UK, 2005.
9. Gil, A.; Salgado, L.; Galicia, L.; Gonzales, I. Predominance-zone diagrams of Fe(III) and Fe(II) sulfate complexes in acidic media. *Talanta* **1995**, *42*, 407–414. [CrossRef]
10. Borra, C.R.; Pontikes, Y.; Binnemans, K.T.; Gerven, V. Leaching of rare earths from bauxite residue (red mud). *Miner. Eng.* **2015**, *76*, 20–27. [CrossRef]
11. Sayan, E.; Bayramoglu, M. Statistical modeling of sulfuric acid leaching of TiO_2 from red mud. *Hydrometallurgy* **2004**, *71*, 397–401. [CrossRef]

© 2017 by the authors. Licensee MDPI, Basel, Switzerland. This article is an open access article distributed under the terms and conditions of the Creative Commons Attribution (CC BY) license (http://creativecommons.org/licenses/by/4.0/).

Article

Chemical Degradation of a Mixture of tri-*n*-Octylamine and 1-Tridecanol in the Presence of Chromium(VI) in Acidic Sulfate Media

Alexandre Chagnes [1,2,*] and Gérard Cote [3]

1. GéoRessources—UMR CNRS 7359-CREGU-Université de Lorraine, 2 Rue du Doyen Roubault, 54518 Vandoeuvre-lès-Nancy CEDEX, France
2. French Network of Hydrometallurgy Promethee, GDR CNRS 3749, 2 Rue du Doyen Roubault, 54518 Vandoeuvre-lès-Nancy CEDEX, France
3. CNRS, Institut de Recherche de Chimie Paris—PSL Research University, Chimie ParisTech, 11 rue Pierre et Marie Curie, 75005 Paris, France; gerard.cote@chimie-paristech.fr
* Correspondence: alexandre.chagnes@univ-lorraine.fr; Tel.: +33-(0)372-744-544

Received: 9 December 2017; Accepted: 13 January 2018; Published: 15 January 2018

Abstract: The chemical degradation of an extraction solvent composed of a mixture of tri-*n*-octylamine (extractant) and 1-tridecanol (phase modifier) in *n*-dodecane in contact with an acidic aqueous sulfate solution containing chromium(VI) has been investigated. The kinetics of degradation and the degradation products have been determined. GC-MS analyses evidenced the formation of 1-tridecanal, di-*n*-octylamine, N,N,N-octen-1-yl-dioctylamine, and an unidentified degradation compound, which may have contained a double bond and a carboxylic acid function. The mechanisms of degradation have been discussed on the basis of these identified degradation compounds. The study of the degradation kinetics showed that an increase of tri-*n*-octylamine concentration in the organic phase is responsible for a decrease of the degradation rate, while an increase in sulfuric acid concentration in the aqueous phase leads to a strong increase in the degradation rate.

Keywords: Alamine® 336; solvent extraction; chromium(VI); degradation; tri-*n*-octylamine

1. Introduction

Solvent extraction is a well-established technology used in hydrometallurgy to produce high-grade metals such as copper, nickel, cobalt, rare earths, uranium, etc. [1]. The chemistry involved in metal extraction is complex, as redox, hydrolysis, precipitation, acido-basic and complexation equilibria occur both in organic and aqueous phases. Solvent extraction processes are sensitive to the nature of the ores, the redox potential, the pH of the feed solution, etc., and a modification of these parameters can be responsible for a dysfunction of extraction plants, including crud formation, chemical degradation of the extraction solvent, and drops in extraction efficiency and selectivity [2].

Although chemical degradation and crud formation are normal in solvent extraction plants, only few papers have reported these phenomena and thoroughly investigated their origins. Diluents, especially alkanes, are very stable materials, whereas extractants or phase modifiers exhibit a stronger tendency to degrade [3]. For instance, degradation of the organic phases due to the presence of NOx has been observed in the extraction circuits of hydrometallurgical plants [4]. Since the mid-1970s, there have been a few episodes where the tertiary amine has been quickly and severely degraded under certain conditions. In particular, Feather et al. [5] mentioned the chemical degradation of Alamine® 336 (mainly constituted of trioctylamine) in kerosene modified with isodecanol during the recovery of uranium from South African ores containing nitrate from residual amounts of the blasting agent. The leach liquor, which contained nitrate in addition to the uranyl sulfate, was passed through an ion-exchange column to recover uranium. In this process, both the uranium and nitrate were collected by the ion-exchange column. The column was then stripped with strong sulfuric acid, which removed both the uranium and the nitrate. The uranium was then extracted from the ion-exchange eluate with Alamine® 336 in a solvent extraction process. Due to the recycling of the solution, the nitrate concentration built up in the system to a point where an attack on the amine occurred, resulting in the formation of nitrosoamines and other degradation products. More recently, Chagnes et al. [6] showed that tri-*n*-octylamine in *n*-dodecane modified with 1-tridecanol is sensitive to oxidation in the presence of Vanadium (V(V)) initially present in the feed solution. It was shown that the presence of V(V) co-extracted in the extraction solvent, likely as polyvanadates [7], and molecular dioxygen are responsible for the chemical degradation of the extraction solvent via two simultaneous routes: (i) the oxidation of 1-tridecanol by V(V) with formation of radicals that induce, in turn, a series of subsequent degradation reactions, including the degradation of tri-*n*-octylamine into N,N,N-octen-1-yl-dioctylamine and di-*n*-octylamine; and (ii) the oxidation of tri-*n*-octylamine by molecular oxygen catalyzed by extracted V(V) [6]. Such chemical degradations of extraction solvents are of great concern in liquid-liquid extraction processes, because they are responsible for an increase of the operating costs due to dramatic solvent consumption and crud formation [8].

Cr(VI) is also present in the feed solution treated by the solvent extraction process implemented in the Niger plant for uranium production, and this metal species is also known to be a strong oxidant that can oxidize organic molecules such as alcohols, aldehydes or ketones [9–13]. Therefore, it is also of great interest to investigate the oxidation properties of Cr(VI) towards the extraction solvent used in the Niger plant, namely tri-*n*-octylamine in *n*-dodecane modified with 1-tridecanol. In the present work, the ageing of a mixture of tri-*n*-octylamine (extractant) and 1-tridecanol (phase modifier) in *n*–dodecane has been investigated in the presence of a model aqueous sulfuric acid solution containing Cr(VI) dissolved in sulfuric acid.

2. Materials and Methods

2.1. Reagents

n-Dodecane (purity > 99%, Aldrich, Lyon, France), 1-tridecanol (purity > 98%, Aldrich), tri-*n*-octylamine (purity > 99%, Aldrich), di-*n*-octylamine (purity > 99%, Aldrich), *n*-heptane (analytical grade, Aldrich), and ethyl ethanolate (analytical grade, Aldrich) were used as delivered. Sulfuric acid (purity 98%), chromium(VI) oxide (purity > 99.9%) from Aldrich and water (resistivity > 18 MΩ·cm) purified with a milli-Q Gradient system from Millipore Corporation (Molsheim, France) were used for preparing the aqueous solutions.

The concentrations of the solutes were given in the molality scale (mole of solute per kg of *n*-dodecane for organic phases and mole of solute per kg of water for aqueous phases).

2.2. Methods and Equipments

The chemical degradation of the organic phases in contact with sulfuric acid containing Cr(VI) arises from a redox reaction leading to the appearance of Cr(III) in the aqueous phase. Therefore, the oxidation of the extraction solvent was investigated by following the evolution of Cr(III) concentration in the aqueous phase by UV-visible spectroscopy at 417 and 587 nm [10].

tri-*n*-Octylamine dissolved in *n*-dodecane modified by 1-tridecanol was pre-equilibrated with sulfuric acid at 1 mol·kg^{-1} before studying the degradation of the solvent (1-tridecanol is a phase modifier used in the formulation of extraction solvent in order to avoid third-phase formation). Each organic phase (5 mL) was mixed with an aqueous phase (5 mL) containing sulfuric acid and Cr(VI). During the degradation tests, the two phases were stirred at 25 °C with a mechanical stirring apparatus (Gherardt Laboshake, Richwiller, France) thermostated with a Gherard Thermoshake. All samples were protected from the light to avoid photo-induced degradation of the organic compounds in the presence of Cr(VI). The degradation occurred over a period ranging from 1 day and 18 days, depending on the composition of the aqueous phase, and it should be pointed out that the degradation of the organic phases was observed only in the presence of Cr(VI) in the aqueous phases.

Samples of the aqueous phase were periodically removed with an Eppendorf pipette and analyzed by UV-Visible spectroscopy (Agilent Varian, Grenoble, France) before being returned to the experimental flask. At the end of the degradation tests, the aqueous and organic phases were analyzed by GC-MS (Agilent Varian, Grenoble, France).

Before GC-MS analyses, each sample was treated as follows: Aqueous phases were contacted with (i) n-heptane after neutralization with sodium hydroxide to extract apolar degradation products, and then with (ii) ethyl ethanoate to extract polar degradation products; and n-Heptane and ethyl ethanoate phases were dried with magnesium sulfate before GC-MS analyses. Identification of the degradation compounds was performed based on analyses of the mass spectra with the NIST database.

UV-Visible analyses of the aqueous and organic phases were performed with a Carry 100 Scan UV-Vis spectrometer (Agilent Varian, Grenoble, France).

GC-MS analyses were performed with a Varian 3300 gas chromatograph coupled with a Ribermag R1010 C (Nermag S.A., Rueil Malmaison, France) mass spectrometer. A fused silica capillary column (30 m × 0.2 mm × 0.2 µm) with cross-linked octadecanyl silicone of CP Sil 5 CB Low Bleed type (Chrompack, Agilent Varian, Grenoble, France) was used in gas chromatography. For gas chromatography experiments, the detector temperature was equal to 250 °C, the injection was of 1 µL in splitless mode, the carrier gas was helium, and the ionizing energy was equal to 70 eV.

3. Results

3.1. Identification of Degradation Products

The extraction solvent composed of tri-n-octylamine, 1-tridecanol and n-dodecane was loaded with Cr(VI) at 25 °C by contacting it with an acidic sulfate solution containing Cr(VI). Both phases were kept in contact under shaking for 15 h at 25 °C. During the degradation tests, gas chromatography-mass spectrometry (GS-MS) showed the presence of a series of degradation compounds, namely 1-tridecanal, di-n-octylamine, n-dodecane tridecanoate and N,N,N-octen-1-yl-dioctylamine, in addition to the starting compounds of the extraction solvent (1-tridecanol, tri-n-octylamine and n-dodecane). The retention times of these compounds are gathered in Table 1.

Table 1. Retention time of the products identified by GC-MS analyses in the extraction solvent after its degradation.

Products	Retention Time/min
n-dodecane	6.20
1-tridecanal	9.10
1-tridecanol	9.60
Unidentified compound	10.40
n-dodecane tridecanoate	11.30
tri-n-Octylamine	20.00
di-n-Octylamine	20.60
N,N,N-octen-1-yl-dioctylamine	29.50

An unidentified degradation compound that may contain a double bond and a carboxylic acid function was also identified, but it is difficult to draw conclusions about its exact nature. The details of the mass spectra have been reported elsewhere [6].

3.2. Degradation Mechanisms

The diagram of speciation of Cr(VI) was calculated using the free software Medusa with the thermodynamic constants reported in Table 2 [14], and shows that, below pH 6, Cr(VI) exists mainly as $HCrO_4^-$ and $Cr_2O_7^{2-}$, which are well-known oxidant species (Figure 1) [12].

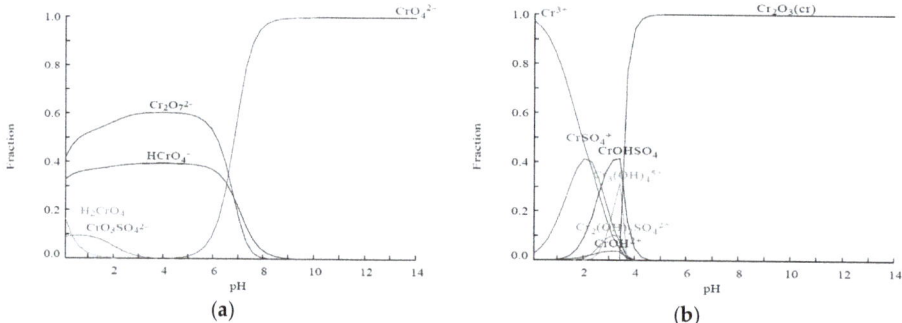

Figure 1. Speciation diagrams of (a) Cr (VI) and (b) Cr(III). [Cr] = 0.05 mol·L^{-1}, ionic strength = 0.1 mol·L^{-1}, T = 25 °C. Calculation performed with Medusa by using the thermodynamic data reported in Table 2.

Table 2. Thermodynamic constants (25 °C) for the calculation of the speciation diagram of Cr (VI) [14].

Equilibria	Thermodynamic Constants at 25 °C
$CrO_4^{2-} + H^+ = HCrO_4^-$	3.5×10^6
$CrO_4^{2-} + 2H^+ = H_2CrO_4$	2.2×10^6
$2CrO_4^{2-} + 2H^+ = Cr_2O_7^{2-} + H_2O$	4.9×10^{14}
$CrO_4^{2-} + 2H^+ + SO_4^{2-} = CrO_3SO_4^{2-} + H_2O$	9.9×10^8
$Cr^{3+} + SO_4^{2-} = CrSO_4^+$	2.2×10^1
$Cr^{3+} + SO_4^{2-} + H_2O = H^+ + CrOHSO_4$	4.5×10^{-2}
$3Cr^{3+} + 4H_2O = 4H^+ + Cr_3(OH)_4^{5+}$	7.1×10^{-9}
$2Cr^{3+} + 2H_2O + SO_4^{2-} = 2H^+ + Cr_2(OH)_2SO_4^{2+}$	8.3×10^{-4}
$Cr^{3+} + H_2O = CrOH^{2+} + H^+$	2.7×10^{-4}

Thus, the degradation mechanisms of tri-*n*-octylamine diluted in *n*-dodecane modified with 1-tridecanol could be explained mainly in consideration of $HCrO_4^-$ and $Cr_2O_7^{2-}$ as the oxidant species.

The UV-Visible spectra of the organic and aqueous phases were recorded throughout the degradation test (Figure 2). Figure 2a shows a peak located at 417 nm in the spectrum of the organic phase attributed to Cr(VI) extracted by tri-*n*-octylamine [10]. The intensity of this peak decreased over time. Figure 2b shows a peak located at 587 nm in the spectrum of the aqueous phase corresponding to Cr(III) [10]. The intensity of this peak increased over time. The decrease of Cr(VI) concentration in the organic phase and the increase of Cr(III) concentration in the aqueous phase can be attributed to the Cr(VI) being stripped from the organic phase into the aqueous phase due to redox reactions responsible for the reduction of Cr(VI) into Cr(III). Cr(III) was spontaneously stripped from the organic phase into the aqueous phase likely due to its having a weak affinity with the extraction solvent because of the electrostatic repulsion between Cr(III) and protonated tri-*n*-octylamine (see speciation diagram of Cr(III) in Figure 1b, which shows that Cr(III) exists as a cation species).

The reduction of Cr(VI) into Cr(III) can be described by the following mechanisms [15]:

$$Cr(VI)_{org} + S_{org} \underset{k_1}{\overset{k_{-1}}{\rightleftarrows}} C^*_{org} \overset{k_2}{\rightarrow} Cr(III)_{org} + P_{org}$$
$$\updownarrow$$
$$Cr(III)_{aq} \tag{1}$$

where C* is a transition complex, S is the organic compound that undergoes oxidation, P is the oxidation product and the subscripts "aq" and "org" denote the aqueous and organic phases, respectively.

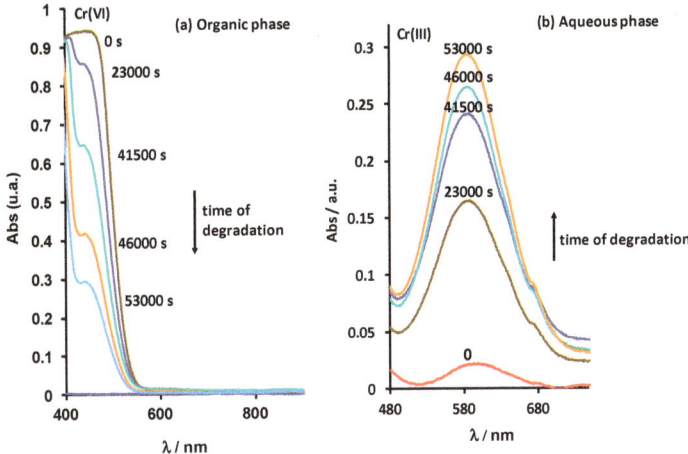

Figure 2. (a) UV-Visible spectra of the organic phase throughout the degradation (organic phase contained initially 0.2 mol·kg^{-1} tri-n-octylamine diluted in n-dodecane modified with 5 wt % 1-tridecanol + 0.05 mol·kg^{-1} Cr(VI); (b) UV-Visible spectra of the aqueous phase in contact with the organic phase throughout the chemical degradation of the organic phase. Temperature = 25 °C.

The net 3e$^-$ reduction of Cr(VI) to Cr(III) may proceed in different ways through the formation of various intermediates like Cr(V) and Cr(IV). The mechanism path of the reduction depends on the nature of the reductant and the reaction conditions [16,17]. Watanabe and Westheimer [18] proposed the following mechanism for the reduction of Cr(VI) to Cr(III) by the substrate S (e.g., alcohol, aldehyde, ketones, etc.) as the two-equivalent reductant:

$$Cr(VI) + S \leftrightarrows Cr(IV)\text{-}S \tag{2a}$$

$$Cr(VI)\text{-}S \rightarrow Cr(IV) + P \tag{2b}$$

$$Cr(IV) + Cr(VI) \rightarrow 2\,Cr(V) \tag{2c}$$

$$Cr(V) + S \rightarrow P + Cr(III) \tag{2d}$$

The following oxidation mechanism can also occur via the formation of radical species, as suggested by Rocek [19]:

$$Cr(VI) + S \leftrightarrows Cr(VI)\text{-}S \tag{2a}$$

$$Cr(VI)\text{-}S \rightarrow Cr(IV) + P \tag{2b}$$

$$Cr(VI)\text{-}S \rightarrow Cr(III) + R\bullet \tag{3a}$$

$$Cr(VI)\text{-}S + R\bullet \rightarrow Cr(V) + P \tag{3b}$$

$$Cr(V) + S \rightarrow Cr(III) + P \tag{3c}$$

The chemical degradation of tri-n-octylamine and 1-tridecanol in n-dodecane in the presence of Cr(VI) dissolved in sulfuric acid likely occurs via the mechanism reported in Equations (2a)–(2d), since no free radicals were observed in the acrylonitrile test (the absence of free radicals is evidenced by no polymerization having occurred after adding a few drops of acrylonitrile to the media) and only Cr(III) species was evidenced in the organic and aqueous phases by UV-visible spectroscopy. The absence of the Cr(V) signature at 510 nm in the UV-visible spectra during chemical degradation may be due to the fast and full reduction of Cr(V) to Cr(III), as reported in Equation (2d) [20].

The oxidation of 1-tridecanol by Cr(VI) can lead to the formation of 1-tridecanal and Cr(IV), as reported in Scheme 1 [12]. In this scheme, the oxidation of 1-tridecanol (RCH_2OH) by $HCrO_4^-$ is represented, and the same mechanism could be considered with $Cr_2O_7^{2-}$.

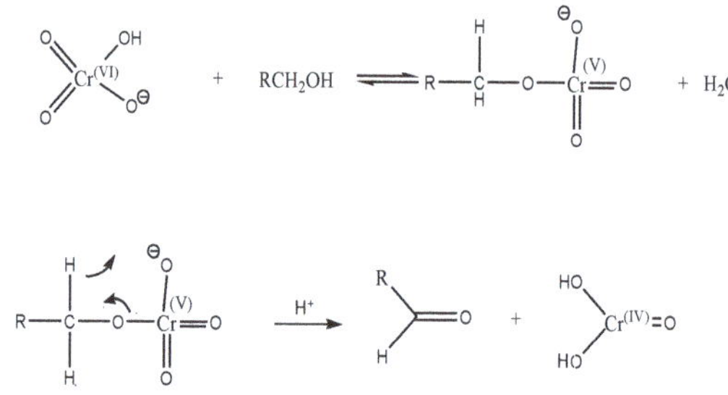

Scheme 1. Alcohol oxidation mechanism.

Afterwards, Cr(IV) reacts with Cr(VI) to form Cr(V) (Equation (2c)), which can be reduced into Cr(III) during the oxidation of 1-tridecanol into 1-tridecanal (Equation (2d)). As Cr(III) has no affinity with tri-n-octylamine, Cr(III) is stripped from the organic phase into the aqueous phase.

Likewise, n-dodecane tridecanoate can be formed by the reaction between 1-tridecanol and 1-tridecanal in the presence of Cr(V) or Cr(VI) as depicted in Scheme 2 [21–24]:

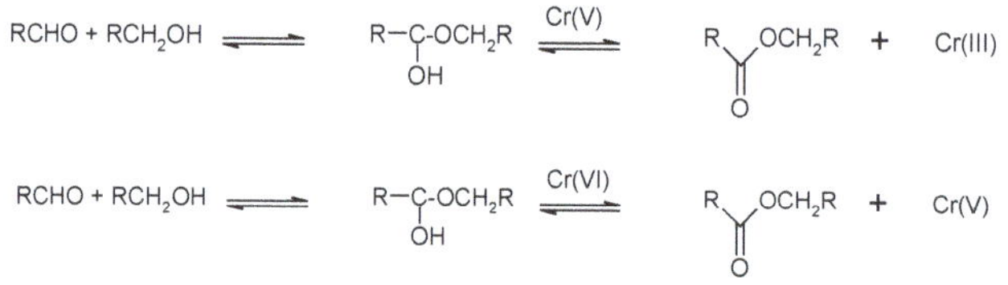

Scheme 2. Aldehyde oxidation mechanisms (R = tridecyl).

A possible mechanistic scenario to account the formation of secondary amines as degradation products is presented in Scheme 3, where R_3NH^+ is the protonated form of TOA (org and aq denote the organic phase and the aqueous phase, respectively) [25]:

- TOA protonation: $(R_3N)_{org} + (H_2SO_4)_{aq} \rightarrow [(R_3NH)^+)_2 \cdot SO_4^{2-}]_{org}$
- Anion exchange between SO_4^{2-} and $HCrO_4^-$ ($HCrO_4^-$ extraction) [26]:
- $[(R_3NH)^+)_2 \cdot SO_4^{2-}]_{org} + 2(HCrO_4^-)_{aq} \rightarrow 2[(R_3NH)^+ \cdot (HCrO_4^-)_2]_{org} + (SO_4^{2-})_{aq}$
- Oxidation reactions (in the organic phase) [25]:

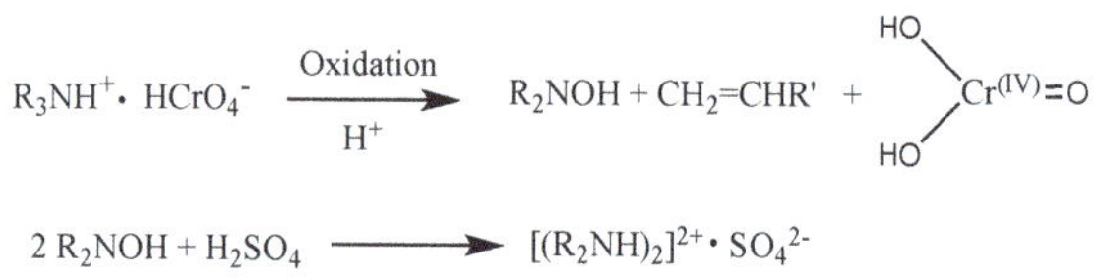

$R_3NH^+ \cdot HCrO_4^- \xrightarrow[H^+]{\text{Oxidation}} R_2NOH + CH_2=CHR' + Cr^{(IV)}(OH)_2=O$

$2 R_2NOH + H_2SO_4 \longrightarrow [(R_2NH)_2]^{2+} \cdot SO_4^{2-}$

Scheme 3. Oxidative mechanism of di-n-octylamine formation.

This scheme involves an initial oxidation of tri-n-octylamine into tri-n-octylamine oxide by Cr(VI). Subsequent degradation of the amine oxide via a 5-membered intermediate generates the protonated dialkylhydroxyamine, which can subsequently undergo N-O bond cleavage via nucleophilic attack at the hydroxyl function [25].

3.3. Kinetics of Degradation

The following empirical equation can be used to describe the kinetics of degradation of trioctylamine-tridecanol in *n*-dodecane by Cr(VI):

$$Ln\left(\frac{A_\infty^{Cr(III)} - A^{Cr(III)}}{A_\infty^{Cr(III)}}\right) = -k_{obs}t \qquad (4)$$

where $A_\infty^{Cr(III)}$ is the absorbance of Cr(III) in the aqueous phase at infinite time (at the end of the oxidation, when the absorbance remains constant), $A^{Cr(III)}$ is the absorbance of Cr(III) in the aqueous phase at the time t and k_{obs} is an empirical kinetic parameter.

The kinetic parameter (k_{obs}) has been deduced by using Equation (4) and the absorbance data of the aqueous phase recorded at 587 nm vs. the time of degradation of the extraction solvent in the presence of Cr(VI). The influence of 1-tridecanol and tri-*n*-octylamine concentrations in the organic phase, and sulfuric acid and chromium (VI) concentrations in the aqueous phase on the kinetics of degradation of the extraction solvent has been reported in Figure 3.

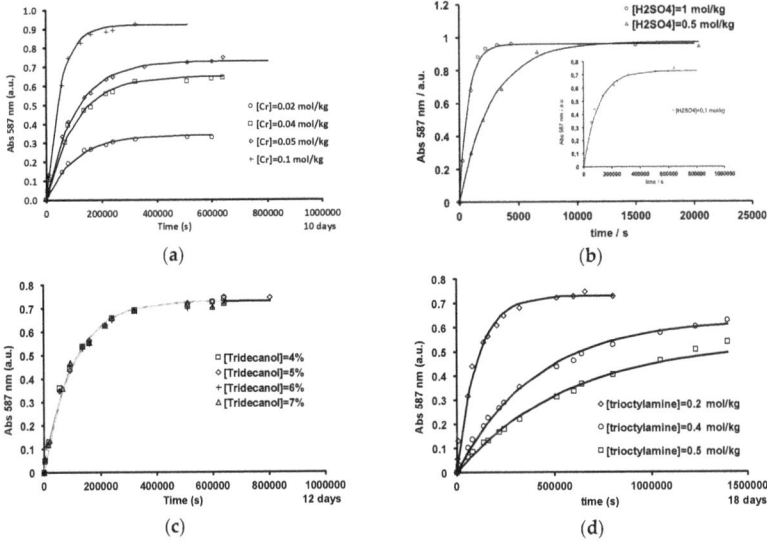

Figure 3. Variation of the absorbance at 587 nm of the aqueous phase during the chemical degradation of 0.2 mol·kg^{-1} tri-*n*-octylamine in *n*-dodecane modified by 1-tridecanol. (**a**) aqueous phase = 0.1 mol·kg^{-1} H$_2$SO$_4$ + Cr(VI) at different concentrations; organic phase: 0.2 mol·kg^{-1} tri-*n*-octylamine + 5 wt % 1-tridecanol; (**b**) aqueous phase = different concentrations of H$_2$SO$_4$ + 0.05 mol·kg^{-1} Cr(VI); organic phase: 0.2 mol·kg^{-1} tri-*n*-octylamine + 5 wt % 1-tridecanol; (**c**) aqueous phase = 0.1 mol·kg^{-1} H$_2$SO$_4$ + 0.05 mol·kg^{-1} Cr(VI); organic phase: 0.2 mol·kg^{-1} tri-*n*-octylamine + different concentrations of 1-tridecanol, (**d**): aqueous phase = 0.1 mol·kg^{-1} H$_2$SO$_4$ + 0.05 mol·kg^{-1} Cr(VI); organic phase: different concentrations of tri-*n*-octylamine + 5 wt % 1-tridecanol. Temperature = 25 °C. Data were fitted with Equation (4).

There is good agreement between experimental absorbance at 587 nm and the absorbance calculated with Equation (4), as the correlation coefficients reported in Table 3 range from 0.972 to 0.955. The same results are obtained with the absorbance at 417 nm. The fitting parameters (k_{obs}) calculated from Equation (4) are gathered in Table 3.

Table 3. Kinetic rates of degradation (k_{obs}) at 25 °C for different compositions of the extraction solvent and different concentrations in sulfuric acid and Cr(VI). In parentheses: correlation coefficient.

[H_2SO_4] mol·kg^{-1}	[Cr(VI)] mol·kg^{-1}	[1-Tridecanol] wt %	[tri-n-Octylamine] mol·kg^{-1}	k_{obs} s^{-1}
0.1	0.02	5	0.2	9.7×10^{-6} (0.989)
0.1	0.04	5	0.2	9.5×10^{-6} (0.977)
0.1	0.05	5	0.2	9.6×10^{-6} (0.972)
0.1	1	5	0.2	2.0×10^{-5} (0.985)
1	0.02	5	0.2	1.1×10^{-3} (0.995)
1	0.04	5	0.2	1.1×10^{-3} (0.993)
1	0.05	5	0.2	1.0×10^{-3} (0.994)
1	1	5	0.2	1.0×10^{-3} (0.978)
0.1	0.02	4	0.2	9.6×10^{-6} (0.975)
0.1	0.04	5	0.2	9.6×10^{-6} (0.972)
0.1	0.05	6	0.2	9.6×10^{-6} (0.965)
0.1	1	7	0.2	9.6×10^{-6} (0.961)
0.1	0.05	5	0.2	9.6×10^{-6} (0.972)
0.1	0.05	5	0.4	2.5×10^{-6} (0.992)
0.1	0.05	5	0.5	1.8×10^{-6} (0.993)

Examination of Table 3 shows that the kinetics of degradation is slower when tri-n-octylamine concentration increases, as k_{obs} = 9.6×10^{-6} s^{-1} at 0.2 mol·kg^{-1} whereas k_{obs} = 1.8×10^{-6} s^{-1} at 0.5 mol·kg^{-1} of tri-n-octylamine (Table 3). At 0.1 mol·kg^{-1} H_2SO_4, the kinetic constant remains constant when Cr(VI) concentration is lower than 0.05 mol·kg^{-1} and then increases at higher concentrations. On the other hand, no influence of Cr(VI) concentration in the aqueous phase is observed when H_2SO_4 concentration is equal to 1 mol·kg^{-1} (Table 3).

4. Conclusions

Cr(VI) can degrade tri-n-octylamine diluted in n-dodecane modified with 1-tridecanol. The degradation of this extraction solvent by Cr(VI) led mainly to the formation of 1-tridecanal, n-dodecane tridecanoate and di-n-octylamine. The degradation of the extraction solvent occurred via the oxidation of 1-tridecanol into 1-tridecanal and the degradation of tri-n-octylamine into dialkylhydroxyamine leading to the formation of di-n-octylamine in the presence of Cr(VI).

tri-n-Octylamine concentration, Cr(VI) concentration and sulfuric acid concentration significantly affected the kinetics of degradation. In particular, an increase of the acidity of the aqueous phase and Cr(VI) concentration were responsible for a dramatic increase of degradation kinetics. It is, therefore, important to decrease the acidity of the aqueous solution and to limit the presence of chromium(VI) by playing on the redox of the leaching solution in the hydrometallurgical process.

Acknowledgments: The authors want to acknowledge AREVA/SEPA for their support.

Author Contributions: Alexandre Chagnes wrote the paper and make experiments. Gerard Cote was involved in the discussions.

Conflicts of Interest: The authors declare no conflict of interest.

References

1. Chagnes, A. Fundamentals in Electrochemistry and Hydrometallurgy. In *Lithium Process Chemistry: Resources, Extractions, Batteries and Recycling*, 1st ed.; Chagnes, A., Swiatowska, J., Eds.; Elsevier: Amsterdam, The Netherlands, 2015; pp. 41–80. ISBN 978-0-12-801417-2.
2. Chagnes, A.; Cote, G.; Courtaud, B.; Syna, N.P.; Thiry, J. Influence of the chemical degradation of trioctylamine dissolved in n-dodecane modified with tridecanol on uranium extraction process in a plant located in Niger. In Proceedings of the 3rd Conference on Uranium, 40th Annual Hydrometallurgy Meeting, Saskatoon, SK, Canada, 15–18 August 2010.

3. Rydberg, J. *Solvent Extraction: Principles and Practices*, 2nd ed.; Marcel Dekker Inc.: New York, NY, USA, 2004; p. 750.
4. Munyungano, B.; Feather, A.; Virnig, M. Degradation problems with the solvent extraction organic at Rössing uranium. In Proceedings of the International Solvent Extraction Conference (ISEC 2008)—Solvent Extraction: Fundamentals to Industrial Applications, Tucson, AZ, USA, 15–19 September 2008; Moyer, B., Ed.; Canadian Institute of Mining, Metallurgy and Petroleum: Montreal, QC, Canada, 2008; Volume 1, pp. 269–274.
5. Feather, A.; Virnig, M.; Bender, J.; Crane, P. Degradation problems with uranium solvent extraction organic. In Proceedings of the ALTA International Uranium Conference, Melbourne, QC, Canada, 23–27 May 2009; ALTA Metallurgical Services: Perth, Australia, 2009.
6. Chagnes, A.; Fossé, C.; Courtaud, B.; Thiry, J.; Cote, G. Chemical degradation of tri-*n*-octylamine and 1-tridecanol dissolved in *n*-dodecane in solvent extraction processes involving aqueous acidic sulfate solutions containing vanadium(V). *Hydrometallurgy* **2011**, *105*, 328–333. [CrossRef]
7. Chagnes, A.; Rager, M.-N.; Courtaud, B.; Thiry, J.; Cote, G. Speciation of vanadium (V) extracted from acidic sulfate media by tri-*n*-octylamine in *n*-dodecane modified with 1-tridecanol. *Hydrometallurgy* **2010**, *104*, 20–24. [CrossRef]
8. Collet, S.; Chagnes, A.; Courtaud, B.; Thiry, J.; Cote, G. Solvent Extraction of Uranium from Acidic Sulfate Media by Alamine® 336: Computer Simulation and Optimization of the Flowsheets. *J. Chem. Technol. Biotechnol.* **2009**, *84*, 1331–1337. [CrossRef]
9. Best, P.; Littler, J.S.; Waters, W.A. The mechanism of oxidation of cyclohexanone under acid conditions. Part 1. Two-electron oxidants. *J. Chem. Soc.* **1962**, 822. [CrossRef]
10. Hartford, W.H.; Darrin, M. The chemistry of chromyl compounds. *Chem. Rev.* **1958**, *58*, 1–61. [CrossRef]
11. Rocek, J.; Riehl, A. Mechanism of the chromic acid oxidation of ketones. *J. Am. Chem. Soc.* **1967**, *89*, 6691–6695. [CrossRef]
12. Stewart, R. *Oxidation Mechanisms—Application to Organic Chemistry*; Benjamin, W.A., Ed.; Wiley: New York, NY, USA, 1964; pp. 33–57.
13. Wagner, R.B.; Moore, J.A. The rearrangement of α,α'-dibromoketones. *J. Am. Chem. Soc.* **1950**, *72*, 974–977. [CrossRef]
14. Puigdomenech, I. *Medusa Software*; KTH University: Stockholm, Sweden, 2000.
15. Da Silva, M.F.C.G.; Da Silva, J.A.L.; Da Silva, J.J.R.F.; Pombeiro, A.J.L.; Amatore, C.; Verpeaux, J.-N. Evidence for a Michaelis-Menten type mechanism in the electrocatalytic oxidation of mercaptopropionic acid by an amavadine model. *J. Am. Chem. Soc.* **1996**, *118*, 7568–7573. [CrossRef]
16. Katz, S.A.; Salem, H. *The Biological Environmental Chemistry of Chromium*; VCH Publishers Inc.: New York, NY, USA, 1994; p. 65.
17. O'Brien, P.; Kotenkamp, A. The chemistry underlying chromate toxicity. *Transit. Met. Chem.* **1995**, *20*, 636–642. [CrossRef]
18. Watanabe, W.; Westheimer, F.H. The kinetics of the chromic acid oxidation of isopropyl alcohol: The induced oxidation of manganous ion. *J. Chem. Phys.* **1949**, *17*, 61–70. [CrossRef]
19. Rocek, J.; Radkowsky, A.E. Mechanism of the chromic acid oxidation of cyclobutanol. *J. Am. Chem. Soc.* **1973**, *95*, 7123–7132. [CrossRef]
20. Sreelatha, G.; Rao, M.P. Kinetics and mechanism of oxidation of allyl, crotyl and cinnamyl alcohol by chromium (V). *Transit. Met. Chem.* **1990**, *15*, 31–33. [CrossRef]
21. Patel, S.; Mishra, B.K. Oxidation of Alcohol by Lipopathic Cr(VI): A Mechanistic Study. *J. Org. Chem.* **2006**, *71*, 6759–6766. [CrossRef] [PubMed]
22. Murashashi, S.I. Synthetic Aspects of Metal-Catalyzed Oxidations of Amines and Related Reactions. *Angew. Chem. Int. Ed.* **1995**, *34*, 2443–2465. [CrossRef]
23. Sheldon, R.A.; Kochi, J.K. *Metal-Catalysed Oxidations of Organic Compounds: Mechanistic Principles and Synthetics Methodology Including Biomedical Processes*; Academic Press: New York, NY, USA, 1981; Chapter 13; pp. 387–397.
24. Velusamy, S.; Punniyamurthy, T. Novel Vanadium-Catalyzed Oxidation of Alcohols to Aldehydes and Ketones under Atmospheric Oxygen. *Org. Lett.* **2004**, *6*, 217–219. [CrossRef] [PubMed]

25. Edwards, C.R.; Garratt, D.G.; De Cesare, J.M.; Olivier, A.J. Solvent Degradation and High Acid Stripping in the Rabbit Lake Uranium Mill. In Proceedings of the SME Annual Meeting, Society of Mining Engineers, Phoenix, AZ, USA, 25–28 January 1988; pp. 1–18.
26. Yaou-Huei, H.; Chen, C.-Y.; J-Kuo, J.-F. Chromium(VI) complexation with triisooctylamine in organic solvents. *Bull. Chem. Soc. Jpn.* **1991**, *64*, 3059–3062.

© 2018 by the authors. Licensee MDPI, Basel, Switzerland. This article is an open access article distributed under the terms and conditions of the Creative Commons Attribution (CC BY) license (http://creativecommons.org/licenses/by/4.0/).

Article

Recovery of Metals from Secondary Raw Materials by Coupled Electroleaching and Electrodeposition in Aqueous or Ionic Liquid Media

Nathalie Leclerc [1,2], Sophie Legeai [1,2], Maxime Balva [1], Claire Hazotte [3], Julien Comel [1], François Lapicque [2,3], Emmanuel Billy [2,4] and Eric Meux [1,2,*]

1. Groupe Chimie et Electrochimie des Matériaux, Institut Jean Lamour, CNRS—Université de Lorraine, 1 Boulevard Arago, BP 95823, 57078 Metz CEDEX 3, France; nathalie.leclerc@univ-lorraine.fr (N.L.); sophie.legeai@univ-lorraine.fr (S.L.); maxime.balva@univ-lorraine.fr (M.B.); julien.comel@univ-lorraine.fr (J.C.)
2. French Network of Hydrometallurgy Promethée, GDR CNRS 3749, 2 rue du Doyen Roubault, 54518 Vandoeuvre-lès-Nancy, France; francois.lapicque@univ-lorraine.fr (F.L.); EMMANUEL.BILLY@cea.fr (E.B.)
3. Laboratoire Réactions et Génie des Procédés-CNRS-Université de Lorraine ENSIC—1 rue Grandville, 54001 Nancy, France; Claire.Hazotte@econick.fr
4. Laboratoire d'Innovation pour les Technologies des Energies nouvelles et les Nanomatériaux, Université de Grenoble-Alpes—CEA, 38054 Grenoble, France
* Correspondence: eric.meux@univ-lorraine.fr

Received: 15 June 2018; Accepted: 17 July 2018; Published: 20 July 2018

Abstract: This paper presents recent views on a hybrid process for beneficiation of secondary raw materials by combined electroleaching of targeted metals and electrodeposition. On the basis of several case studies with aqueous solutions or in ionic liquid media, the paper describes the potential and the limits of the novel, hybrid technique, together with the methodology employed, combining determination of speciation, physical chemistry, electrochemistry, and chemical engineering. On one hand, the case of electroleaching/electrodeposition (E/E) process in aqueous media, although often investigated at the bench scale, appears nevertheless relatively mature, because of the developed methodology, and the appreciable current density allowed, and so it can be used to successfully treat electrode materials of spent Zn/MnO_2 batteries or Ni/Cd accumulators and Waelz oxide. On the other hand, the use of ionic liquids as promising media for the recovery of various metals can be considered for other types of wastes, as shown here for the case of electrodes of aged fuel cells. The combined (E/E) technique could be successfully used for the above waste, in particular by the tricky selection of ionic liquid media. Nevertheless, further investigations in physical chemistry and chemical engineering appear necessary for possible developments of larger-scale processes for the recovery of these strategic resources.

Keywords: metal recovery; electroleaching; electrodeposition; secondary raw materials; ionic liquids

---

1. Introduction

Leaching can be considered to be the key step for most hydrometallurgical processes. This unit operation is classically performed by chemical reactions on the targeted species using various reagents exhibiting either acid-base properties (e.g., leaching of bauxites by soda in the Bayer Process, leaching of ilmenite $FeTiO_3$ with concentrated H_2SO_4 for TiO_2 production), or chelating properties as in the cyanidation of gold ores, or oxidative or reductive properties, as for the oxidation of metal sulfides with Fe^{3+}, or the reduction of MnO_2 with glucose.

To improve its effectiveness, chemical leaching can be performed in an autoclave or can be assisted by the means of ultrasound. In addition to chemical leaching, two other techniques can be considered for ore treatment, namely (i) bioleaching (or bio-oxidation), in which bacteria or fungi are employed to extract metal species; (ii) electroleaching, for which oxidative/reductive reagents are replaced by electrons in electrochemical reactions. Electroleaching is considered to be "direct" when compounds react at the electrodes or "indirect"—or assisted—in case where chemical reagents are generated at the electrodes to perform leaching: an example is the action of anodically generated chlorine from the chloride-based electrolyte.

The first studies dealing with electroleaching in extractive metallurgy were devoted to the treatment of sulfides ores [1]. Due to their good electrical conductivity, metal sulfides were treated by direct electroleaching using consumable anodes formed by the sulfide to be converted, or blended with graphite to improve the conductivity of the formed anodes. Various sulfides have been investigated: chalcocite Cu_2S, chalcopyrite $CuFeS_2$, galena PbS, and sphalerite ZnS. For a divalent sulfide, the anodic dissolution can be written as follows:

$$MS_{(s)} \rightarrow M^{2+} + S_{(s)} + 2\ e^- \tag{1}$$

For the above-mentioned treatments reported in the literature, a simultaneous cathodic deposition occurred.

Kirk et al. [2] investigated the electrochemical leaching of silver arsenopyrite. However, in this case, electroleaching was indirect and consisted of the anodic generation of chlorine gas from hydrochloric acid. Although less encountered, reductive electroleaching of ores can also be considered, as shown by a couple of published papers. This reductive electroleaching, mainly direct, was performed with natural resources containing MnO_2—in which the manganese cation was cathodically dissolved from the oxide present in the low-grade manganese ores [3], or in polymetallic ocean nodules [4]. Apart from the beneficiation of mineral ores, electroleaching can be applied to the treatment of secondary raw materials, in particular, that of printed circuit boards, either by direct electro-dissolution [5,6] or by generation of an oxidizing reagent at the anode [7].

In hydrometallurgical processes, the metals of interest in the form of single or complexed ions are recovered by electrowinning. Therefore, it is possible to imagine hybrid processes where, in a single cell, metals—under various chemical forms—are leached (directly or not) at the anode, the metal ions produced will migrate to the cathode in the same cell, where they are reduced to form the metal deposit. This combined technique is presented more in detail below:

- Coupling electro-assisted leaching to electrodeposition

The principle is described in Figure 1a. The material to be treated is placed in the anodic compartment of the cell. The leaching reagent is produced at the anode (H^+, Cl_2, Fe^{3+} ...). These compounds allow the occurrence of the dissolution of metal species present in the solid-liquid matrix (pulp) under various mineralogical forms e.g., metal, oxides or sulfides. The metal cations formed migrate to the cathode where the reduction can be selective or not. A separator is generally placed between the anode and cathode to prevent the attrition of the metal deposit by the solid particles.

- Coupling direct electroleaching to electrodeposition

The principle is presented in Figure 1b. In this case, the anode is formed by the material to be leached, alone, if its conductivity is sufficient, or mixed with fine carbon powder for enhanced electrical conductivity of the anode.

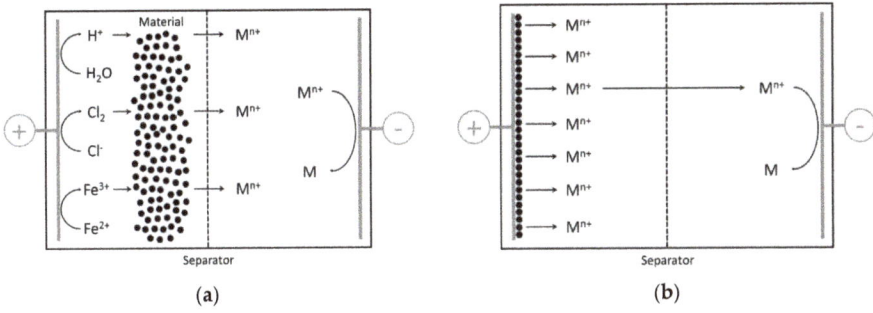

Figure 1. Scheme of the coupled electro-assisted leaching electrodeposition (**a**) and direct electroleaching/electrodeposition (**b**) conducted in a single-cell.

Most studies devoted to coupling electroleaching to electrodeposition in a single-cell have been carried out in aqueous media e.g., H_2SO_4 or HCl solutions, with or without metal salts. The use of non-aqueous media such as ionic liquids (ILs) usually exhibiting broad electrochemical windows, could allow extension of the process to the case of noble metals (Pt, Pd, Rh ...) and to the recovery of metal elements such as rare-earths or tantalum, which is not possible in water-based liquids [8,9].

Examples of electroleaching/electroreduction processes at bench scale are presented in Table 1.

Table 1. Typical examples of coupled electroleaching/electrodeposition in a single-cell.

Materials	Electroleaching	Number of Compartments	Anode with (Reaction)	Cathode with (Reaction)	Electrolyte	Separator	Ref.
Fe/Ni sulfides concentrates	Assisted	2	Graphite (Cl_2 generated by oxidation of Cl^-)	Titanium (Ni deposition)	Anolyte: NaCl/HCl Catholyte: NaCl/HBO$_2$	Cation Exchange Membrane	[10]
ZnO dispersed in a sand matrix	Assisted	3	Lead (H^+ generated by oxidation of water)	Stainless steel (Zn deposition)	Anolyte: H_2SO_4 Catholyte: H_2SO_4/ZnSO$_4$	Polypropylene cloth	[11]
PCB powder	Direct	1	Stainless steel basket	Stainless steel (Cu deposition)	H_2SO_4/CuSO$_4$	-	[12]
Ni/Cd spent batteries	Assisted	2	Platinized titanium (H^+ generated by oxidation of water)	Aluminium (Cd deposition)	Anolyte: diluted H_2SO_4 Catholyte: H_2SO_4/CdSO$_4$	Polypropylene cloth	[13]
Membrane Electrode Assembly of spent PEMFC	Direct	1	Spent MEA	Graphite (Pt deposition)	Ionic liquid mixtures [BMIm]Cl^1 [BMIm]TFSI^2	-	[14]
Ag$_2$S/Ag	Direct	1	Carbon Paste Electrode with Ag$_2$S/Ag	Graphite (Ag deposition)	Na$_2$S$_2$O$_3$ Na$_2$SO$_3$	-	[15]

BMIm = 1-butyl, 3-methylimidazolium; TFSI = bis(trifluorosulfonyl)imide.

This paper is aimed at presenting the potential of metal recovery from secondary raw materials by coupled electroleaching electrodeposition on the basis of the systems shown in Table 1: in spite of the differences in the examples and beyond the performance attained, efforts have been put here on methodological aspects to be accounted for in the design of such processes, namely comparison of the importance of reaction and transfer phenomena, together with the significance of complexation allowed by the solvents, in particular with ionic liquids. The paper is divided into two parts. First, we present coupled electroassisted leaching electroreduction in an aqueous medium for the case of zinc-containing waste issued by steel manufacturing plants [11] and he "black mass" contained in nickel/cadmium spent batteries [13]. In this part, investigation of the various physicochemical phenomena (chemical reaction, transport, electrochemical processes) involved allowed better understanding of the overall process, for possible design and estimation of capacities and limits of the technique. Secondly, the coupled direct electroleaching electrodeposition in ionic liquids has been developed for spent membrane electrode assemblies (MEA) of proton exchange membrane fuel cells (PEMFC) [14], for the recovery of Pt from the electrodes. The works described in this second part are more recent, and the use of ILs for electroleaching/electroreduction is still poorly described in the relevant literature. In this part, emphasis is put more significantly on the physicochemical aspects of Pt species to be dissolved in ionic liquids, which had to exhibit high complexing properties while allowing Pt deposition at the cell cathode.

2. Electro-Assisted Leaching/Electroreduction in Aqueous Media

2.1. Principle

The principle of electro-assisted leaching/electrodeposition studied here consists of the generation of protons by electrochemical oxidation of water according to:

$$2\,H_2O \rightarrow 4\,H^+ + O_{2(g)} + 4\,e^- \qquad (2)$$

Protons produced by reaction (2) can react with metals, metal hydroxides, or metal oxides. Then, M^{z+} cations released, move towards the cathode compartment where they are reduced, selectively or not, with a current efficiency depending on the reversible potential of M^{z+}/M couple and taking into account the possible occurrence of water electro-reduction according to:

$$2\,H_2O + 2\,e^- \rightarrow 2\,OH^- + H_{2(g)} \qquad (3)$$

2.2. Experimental Section and Methodology

2.2.1. Design of the Cell

Tests of electro-assisted leaching coupled to electrodeposition were carried out in a laboratory-made cell [11], made of PVC and consisting of compartments—of a variable number. Figure 2 shows a two-compartment cell. Its detailed description has been presented in previous papers with different configurations in the treatment of zinc-containing waste [16], or black mass from nickel/cadmium spent batteries [17]. In all cases, the cell was assembled with PVC elements (9 cm high and 7 cm wide and with an adjustable depth). Fluid circulation in each compartment was made possible by inlet and outlet tubes: one on the top of one side and a second at the bottom of the other side. A peristaltic pump was used for circulation of the liquids at a flow rate fixed at 0.9 L min^{-1}. The cell—operated batchwise—was not covered by a lid for sampling of the solution, addition, or extraction of the solid waste along the runs, stirring of the solid waste, and monitoring of variables such as pH and potential, and finally for the evacuation of evolved gases e.g., H$_2$. Each compartment was separated from the others by a polypropylene (PP) cloth.

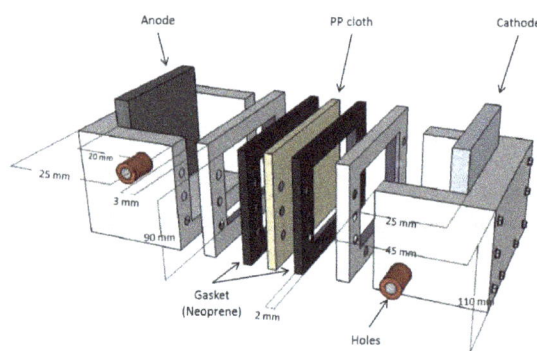

Figure 2. Laboratory-made cell for electro-assisted leaching and electrodeposition in aqueous media.

2.2.2. Solid Waste

Various solids have been investigated in the above cell in the form of ground particles, as follows:

- Synthetic Zn-containing waste: this synthetic solid was prepared by dispersing 10 wt. % zinc oxide ZnO in sand.
- Reconstructed black mass from spent Ni/Cd batteries: the active powders from spent Ni/Cd batteries (so-called "black mass") were used. Black mass was recovered from end-of-life SANYO batteries after manual dismantling [18]. The active powders of anode and cathode were gathered and homogenized to give a reconstructed black mass. This material has the following composition: 51 wt. % $Cd(OH)_2$, 30 wt. % $Ni(OH)_2$, 13 wt. % Ni, approx. 1 wt. % NiOOH and about 2.5 wt. % $Co(OH)_2$.
- Waelz oxide was produced by carbothermal reduction of electric arc furnace dust (EAFD). The sample studied was provided by Recytech S.A (Fouquières-lès-Lens, France). In this sample, zinc was present in different forms: 90% ZnO, 8% Zn and 2% $ZnFe_2O_4$)
- Black mass from spent Zn/MnO_2 batteries. A sample of industrial black mass was provided by Eurodieuze Industrie (Dieuze, France). This material was obtained after the different mechanical stages as follows: crushing of spent batteries, magnetic sorting to recover steel chips, Eddy current sorting to recover non-ferrous metals, and removal of plastics and papers by air entrainment. Zinc and manganese form distributions were as follows: zinc (47.3% ZnO and 52.7% Zn), and manganese (35.4% MnOOH and 64.6% MnO_2).
- Industrial black mass from spent Ni/Cd batteries. This sample was also provided by Eurodieuze Industrie (Dieuze, France). Its composition is given in Table 2 [19]. The main difference with reconstructed black mass is the presence of small amounts of carbon and $Fe/Fe(OH)_3$, resulting from incomplete magnetic sorting.

The three black mass solids investigated consisted of coarse particles, with an average diameter near 1 mm, whereas the particles of zinc-containing solids were somewhat finer, with an average diameter in the order of 200 µm.

Table 2. Composition of industrial black mass coming from spent Ni/Cd batteries [19].

Black Mass Composition (wt. %)	$Ni(OH)_2$	Ni	$Cd(OH)_2$	Cd	$Co(OH)_2$	Co	$Fe(OH)_3$	Fe	C
	28.7	20.2	37.2	3.3	1.6	0.3	1.0	0.6	2.0

2.2.3. Experiments

As a first step and for each waste, leaching and electrodeposition have been separately investigated in conventional laboratory cells. These preliminary investigations aimed at determining the leaching kinetics of the waste components and the optimum current density for obtaining a compact, regular metal deposit.

Then, these two steps were achieved in the cell shown in Figure 2 with various configurations, in particular with two or three compartments, with or without fluid circulation and with or without stirring of the solid-liquid suspension, in order to follow the time variations of the concentrations of the various species. For investigation of electro-assisted leaching without electrodeposition of the leached species, an anionic-supported membrane (Eurodia) was installed to replace the PP cloth, which separated the cathodic compartment from the central chamber containing the suspension [11,20].

Tests of combined electroleaching-electrodeposition were aimed to define the best electrolytes compositions and the optimal operating conditions (current density, influence of fluid circulation, and waste stirring) for the highest performance of the cell, in terms of leaching and deposition efficiencies, quality of the metal product, and purity of the solution recovered at the end of the runs.

- Configuration of the cell for zinc electroleaching/electrodeposition

A three-compartment cell was used (Figure 3). Synthetic waste was introduced in the central chamber (2.5 cm thick) separated from anodic and cathodic compartments by polypropylene clothes (Sefar Fyltis).

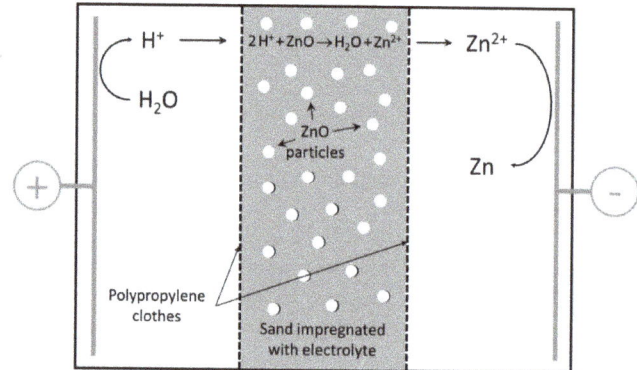

Figure 3. Cell configuration for electro-assisted leaching/electrodeposition applied to ZnO/sand mixture.

A lead plate anode was inserted in the cell, whereas the cathode was either a stainless steel or an aluminum plate. Both electrodes had the same active area, at 53 cm². The anolyte was a diluted H_2SO_4 solution to ensure a sufficient electrical conductivity and to minimize the ohmic drop in the early stages of the experiment. For the same reason, ZnO/sand solid was impregnated by diluted H_2SO_4. The catholyte was a mixed H_2SO_4, $ZnSO_4$ solution.

- Configuration of the cell for Ni/Cd electroleaching/electrodeposition

A two-compartment cell was used (Figure 4) with a platinized titanium anode for O_2 and H^+ generation, and an aluminum cathode (active area near 53 cm²). Black mass was directly added to the anodic compartment with diluted H_2SO_4 as the anolyte solution. Catholyte was a mixed $H_2SO_4/CdSO_4$ solution. A PP cloth (Mortelecque, France) separated the two compartments.

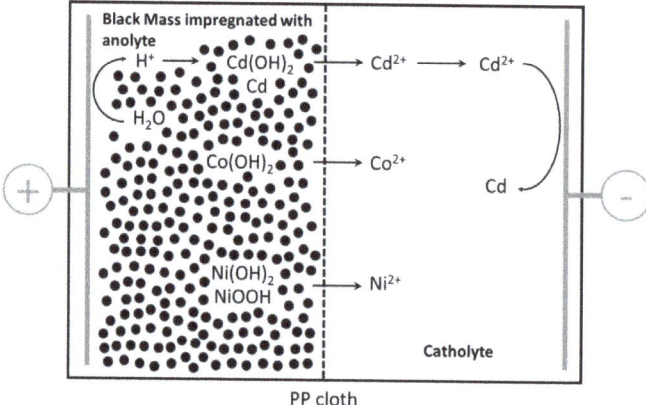

Figure 4. Cell configuration for electroleaching/electrodeposition applied to Ni/Cd Black Mass.

In comparison with the ZnO/sand blend, leaching reactions were more complex. Electro-generated protons were consumed in:

- Acidic dissolution of metal hydroxides after:

$$M(OH)_2 + 2\,H^+ \rightarrow M^{2+} + 2\,H_2O^+ \tag{4}$$

- Oxidation of metal cadmium:

$$Cd + 2\,H^+ \rightarrow Cd^{2+} + H_{2(g)} \tag{5}$$

- Reduction of NiOOH:

$$4\,NiOOH + 8\,H^+ \rightarrow 4\,Ni^{2+} + 6\,H_2O + O_{2(g)} \tag{6}$$

The three industrial samples were investigated with the cell configuration shown in Figure 4.

2.3. Results and Discussion

2.3.1. Synthetic Zinc-Containing Waste

The study of optimal conditions of zinc electrowinning [21] had shown that operating at a current density of 450 A·m^{-2} allowed formation of a compact zinc deposit. The best electrolyte compositions for combined electro-assisted leaching and electrodeposition treatment were found to be as follows [11]: (i) 125 mL of H$_2$SO$_4$ 0.01 M introduced in the cathode chamber; (ii) the central (waste) compartment filled with 100 mL of 0.01 M H$_2$SO$_4$, 0.5 M ZnSO$_4$ in the presence of 133 g ZnO-sand mixture; (iii) the catholyte was 125 mL 0.5 M H$_2$SO$_4$ 0.5 M ZnSO$_4$ medium. This last composition was selected after thorough studies of zinc deposition in a Hull cell with various Zn sulfate sulfuric solutions [21].

In typical runs, the three-compartment cell was operated for 6 h at 450 A·m^{-2}: 97% of the ZnO introduced was leached, and 75% of the leached zinc could be recovered in the form of compact metal without dendrites and exhibiting no powdery aspect—this fraction is called here deposition yield, and the deposition current efficiency attained 60%. It was also shown that circulation of electrolytes and stirring of the pulp had a positive influence on leaching and deposition processes, for the latter process in terms of current efficiency and morphology of the deposit produced.

The power consumption for this treatment was estimated at 23.7 kWh·kg^{-1}. Even if this value was in good agreement with the work of Page et al. [10], it was much higher than that required for the only zinc electrodeposition from sulfate/sulfuric medium, near 3.5 kWh·kg^{-1} [22] or even the free enthalpy difference in a zinc performing Zn deposition and oxygen evolution under standard conditions, at approx. 1.63 kWh·kg^{-1}. It should be noted that the power consumption involved in the electro-assisted leaching operation was not accounted for in [22] and the above thermodynamic energy estimated relies also a 100% current efficiency for Zn deposition. Moreover, the design of the cell for these first tests was far from optimal, with presumably significant ohmic drop due to appreciable electrode gap in the laboratory cell, in addition to the electrode overpotentials. Therefore, in order to decrease the power consumption, it was decided that one compartment would be removed to avoid the resistance induced by the cloth, and to reduce the distance between the two electrodes. All tests with industrial wastes were performed with a two-compartment cell.

2.3.2. Reconstructed Black Mass from Spent Ni/Cd Batteries

The objective of this study was to obtain, on one hand, a solid residue containing only metal nickel and carbon, and on the other hand, metal cadmium of acceptable quality. This study also intended to produce a pure cadmium-free solution of metal salts. A current density of 350 A·m^{-2} was shown to correspond to the best conditions for the combined leaching and formation of a compact cadmium deposit at the cathode [20].

Optimum electrolyte compositions [13] were as follows. The cathode chamber was filled with a suspension prepared with 170 mL of 0.05 M H_2SO_4, with 20 g reconstructed black mass, whereas the anolyte was 156 mL 0.05 M H_2SO_4, and 0.6 M $CdSO_4$. It was also observed that more regular operation was allowed with continuous circulation of the two media in the compartment, with stirring of the suspension in the anodic chamber.

During all experiments, samples were collected regularly in each compartment with a view to monitor the variations of cadmium, nickel, cobalt, and acidic species (H^+ and HSO_4^-) concentrations along time.

After 5 h treatment, the solid residue, the cathodic deposit, and the electrolytes were submitted to analysis: the results are reported in Table 3.

Table 3. Composition of electrolytes and solids after 5 hr treatment at 350 A·m^{-2}.

	Cd	Ni	Co	Ni(OH)$_2$	Cd(OH)$_2$	C
Catholyte (g·L^{-1})	0.3	11.2	0.9			
Anolyte (g·L^{-1})	21.4	15.9	0.9	-	-	-
Deposit (wt. %)	99.7	0.2	0.02			
Solid residue (wt. %)	0	79	0	17	2	2

It can be seen in this Table that the residual solid contained predominantly nickel and that the cadmium deposit was of a high purity (99.7%). The leaching yield of cadmium was close to 99%. Because of its low cadmium content, the catholyte should be valuable in nickel industry after hydroxide precipitation. The obtained anolyte was a mixed Ni Cd solution which could be reused to prepare the catholytic solution for further runs.

As also observed for the treatment of zinc-containing waste, the performance and efficiency of the overall operation was governed by several factors: generation of protons at the anode, dissolution of the particles, transport of metal cations to the cathodic chamber, in particular through the cloth, and electrochemical deposition. As a matter of fact, several phenomena appeared crucial in the overall performance [20]:

- Leaching of solid particles was revealed to be controlled by both mass transfer of protons to the solid surface, and by the rate of surface chemical reaction: use of finer waste particles improved the overall process efficiency.
- Transport of generated metal cations occurs by migration, in addition to possible occurrence of convection and diffusion. The current density is fixed in the chronopotentiometric runs, so metal cations are selectively transported to the cathode if their transference number is high, i.e., if that of protons remains at a low/moderate level: for this reason, the solution pH has to obey a compromise between sufficient fast leaching of metal oxides or hydroxides and a moderate content of acidic ions.
- Such compromise in acid concentration is also important for the electrodeposition step at the cathode, for the sake of little significant side evolution of hydrogen, and to avoid formation of metal hydroxide at the cathode surface, with local pH larger than that in the bulk.
- Transport through the PP cloth can be also rate-controlling, in particular with using materials prepared with woven bundles of thin polymeric fibers and exhibiting a low overall permeability.

These conclusions could guide the definition of operating conditions for treatment of industrial wastes, and help in the design of an electroleaching/electrodeposition process.

2.3.3. Industrial Samples

The experimental conditions and results obtained in the optimized treatment of the three industrial wastes are summarized in Table 4.

The data reported in Table 4 allowed the following conclusions on the process to be drawn:

- Waelz oxide. Initially at 5 wt. %, the residual solid was lead-enriched, with a final fraction at 35% after the run. Lead was in PbSO$_4$ mineralogical form. This solid could be reused in lead metallurgy. Moreover, 99.9% zinc deposit was produced, for possible reuse in galvanization processes. The final anolyte could be used as a catholyte for further runs, after upgrading by concentration adjustment.
- Zn–MnO$_2$ spent batteries. The treated solid contained a slight amount of zinc and its manganese concentration was almost increased twofold. Pure zinc was deposited at the cathode, and the electrolytes could be reused for further treatment runs.
- Ni/Cd spent batteries. Nickel and cadmium hydroxides were entirely leached. Metal nickel and carbon concentrations in the solid recovered were increased fourfold. This could be of use in nickel metallurgy. The Cd cathode deposit does not have a large market value but could be dissolved to prepare new electrolytes. The anolyte could be reused as catholyte for further runs or could be of use in nickel metallurgy after precipitation of the hydroxides.

Table 4. Electro-assisted leaching/electrodeposition of various industrials wastes (or byproduct) in the discontinuous process.

		Waelz Oxide [11]	Zn–MnO$_2$ Spent Batteries [11]	Ni/Cd Spent Batteries [19]
Experimental conditions	J (A·m^{-2})	450		350
	Run duration (h)	6	8	6
	Type of anode	Pb		Platinized Titanium
	Anolyte volume (mL)	130	150	170
	Anolyte composition (M)	H$_2$SO$_4$: 0.1 ZnSO$_4$: 0.2		H$_2$SO$_4$: 0.1 Addition of 3 mL 2 M CdSO$_4$ every 60 min
	Type of cathode	Al		Al
	Catholyte volume (mL)	125	150	220
	Catholyte composition (M)	H$_2$SO$_4$: 0.5 ZnSO$_4$: 0.5	H$_2$SO$_4$: 0.25 ZnSO$_4$: 0.5	H$_2$SO$_4$: 0.5 CdSO$_4$: 0.49
	Weight of processed solid (g)	20	25	30
	Procedure for solid addition	Initial introduction		5 g at initial time, then 5 g added every hour
	Procedure for extraction of the residual solid	Extraction at the end of the run		Extraction every hour before addition of solids
	Procedure for extraction of the deposit			Extraction every hour
Yields (%)	Leaching yield	Zn: 99.1	Zn: 96.3	Ni(OH)$_2$ and Cd(OH)$_2$: 100
	Faradaic yield	61	37	52.6
	Deposition yield	61	75.7	56.8
After electro-assisted leaching/electrodeposition				
	Residual solid composition (wt. %)	Zn: 0.3 Pb: 34.9 Fe: 4.7 Si: 10.6	Mn: 45.2 Zn: 1.4 Fe: 1.5	Ni: 82.8 C: 8.7 Cd: 1.4 Fe: 1.1
	Weight of deposit (g)	10.7	8.6	13.4
	Deposit composition (wt. %)	Zn: 99.9 Pb: 0.06	Zn: 99.8 Cu: 0.09	Cd: 90 Cd(OH)$_2$: 10
	Anolyte composition (M)	H$_2$SO$_4$: 0.42 ZnSO$_4$: 0.36	H$_2$SO$_4$: 0.6 ZnSO$_4$: 0.2 MnSO$_4$: 0.1	H$_2$SO$_4$: 0.54 NiSO$_4$: 0.25 CdSO$_4$: 0.13
	Catholyte composition (M)	H$_2$SO$_4$: 0.02 ZnSO$_4$: 0.04	H$_2$SO$_4$: 0.09 ZnSO$_4$: 0.02 MnSO$_4$: 0.1	H$_2$SO$_4$: 0.7 NiSO$_4$: 0.21 CdSO$_4$: 0.02

3. Direct Electroleaching/Electrodeposition of Platinum in Ionic Liquids

Ionic liquids (ILs), which are molten salts with a melting point below 100 °C, are very attractive electrolytes due to their unique properties: they exhibit in particular very low volatility, good thermal stability, and a wide electrochemical window. This window is usually far larger than 2 V, which makes it possible to perform anodic dissolution and electrodeposition of various metals with a limited ionic liquid degradation [8,23,24]. ILs appear promising for the recovery of hardly reducible metals [9,25,26], or noble metals of troublesome oxidation [27–30]. Moreover, ILs composed of coordinating anions (such as Cl^-, Br^-, SCN^-, $N(CN)_2^-$ ...) are particularly interesting for leaching purposes, since the reachable concentration of the ligand can be as high as 10 M.

3.1. Experimental Section

3.1.1. Chemicals

1-butyl-3-methyl imidazolium chloride (BMIM Cl) (99%) and 1-butyl-3-methyl trifluoromethanesulfonate (BMIM OTf) (98%) were purchased from Iolitec®. Silver trifluoromethanesulfonate (Ag OTf) was purchased from ACROS®. 1-butyl-3-methyl imidazolium bromide (BMIM Br) was prepared according to the procedure described in [31]. 1-butyl-3-methyl imidazolium bis(trifluoromethylsulfonyl)imide (BMIM TFSI) was also prepared as reported in [32].

3.1.2. Electrochemical Experiments

All electrochemical experiments were performed at 100 °C, using a VSP-300 (Biologic®, Claix, France) potentiostat controlled with EC-Lab software and a three-electrode cell. The Ag^{+I}/Ag reference electrode was prepared by immersing an Ag wire (Ø = 1 mm, Alfa Aesar®) in a 10 mM AgOTf solution in BMIM OTf placed in a salt bridge (AL120, CTB Choffel). Voltammetric experiments were performed using glassy carbon or platinum disks (Ø = 3 mm and 2 mm, respectively) as working electrodes, and a glassy carbon disk (Ø = 3 mm) as a counter electrode. The sweep rate was fixed at 20 mV·s^{-1}. The disk electrodes were polished mechanically using 5 µm and 0.3 µm abrasive bands (Radiometer Analytical®) before each electrochemical experiment. Electroleaching experiments were performed in potentiostatic mode using a 1 cm^2 platinum plate (99.99%, AMTS®) or a sample of used PEMFC electrode as the anode. A glassy carbon plate with a surface equal to 1 cm^2 was used as the cathode. The volume of solution used was 8 mL. Unless otherwise stated, the experiments were performed under inert atmosphere, in a MBraun® Labstar MB 10 Compact glove box, with O_2 and H_2O contents below 0.5 ppm.

3.1.3. Determination of Dissolved Platinum Concentration

The amount of leached platinum was determined by gravimetry and atomic absorption spectrometry (AAS). The platinum plate anode was weighed before and after its electrochemical dissolution, after rinsing with distilled water and acetone. A quantification method specific to ILs was developed in a previous work to determine platinum concentration by AAS [32]. All AAS measurements were conducted using a Varian® AA 240 spectrometer, controlled with the SpectrAA© software (Agilent Technologies, Santa Clara, CA, USA).

3.1.4. Leaching of Platinum Nanoparticles from MEAs

PEMFC electrodes were supplied by LITEN-CEA in Grenoble. The electrode ink was prepared by mixing Vulcan XC-72 Carbon with 47.3 wt. % of platinum and 26 wt. % of Nafion solution in the mixed organic solvent. The ink was then deposited on the gas diffusion layer consisting of a macroporous substrate treated with 5% PTFE (Polytetrafluoroethylene) and a microporous layer. The average platinum amount on the used PEMFC electrodes recovered was determined to be close to 75 µg·cm^{-2}. The electrodes were characterized before and after dissolution experiments using a Tescan® Vega 3 SBU

Easy Probe SEM (Scanning Electron Microscopy) with a Bruker® XFlash Detector 410 M controlled by Espris© 1.9 Software (Bruker, Billerica, MA, USA) to perform EDX (Energy Dispersive X-ray) analysis.

3.2. Results and Discussion

3.2.1. Electrochemical Leaching and Electrodeposition of Platinum

We demonstrated previously the feasibility of anodic dissolution of platinum followed by its recovery by electrodeposition at a carbon cathode [32,33], using IL electrolytes at temperatures near 100 °C. The electrolyte was a mixture of two ILs having a common cation, BMIM$^+$, associated with two different anions: Cl$^-$ and TFSI$^-$. BMIM Cl was chosen due to the complexing property of chloride anions for platinum. However, whereas a high chloride concentration favors the leaching of platinum, it inhibits the metal electrodeposition, which means that BMIM Cl cannot be used alone here. In order to reduce the complexing ability of the electrolyte and allow occurrence of Pt electrodeposition, BMIM Cl was used as a solute with another IL—BMIM TFSI—acting as the solvent, since TFSI$^-$ anion is known to be a low coordinating species. Moreover, the use of such mixed ILs instead of pure BMIM Cl exhibits further advantages: the mixture is liquid at room temperature, far less viscous than pure ILs at the temperature of interest, and far less hygroscopic. The optimum electrolyte composition was determined for both electrochemical leaching and the electrodeposition of platinum using a single cell process operated in potentiostatic mode.

Further works were dedicated to a deeper understanding of the electrochemical reactions involved in the IL melt. More specifically, the influence of applied potential on the speciation of platinum and on the efficiency of the leaching step was studied in terms of rate and faradaic yield.

Figure 5 presents the linear voltammetric curves obtained using a platinum disk as a working electrode in BMIM TFSI or BMIM Cl alone, and in the mixed IL, whose composition was found to be optimal for Pt recovery.

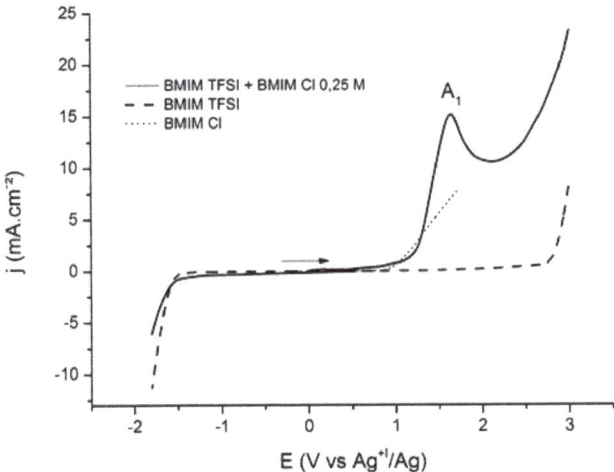

Figure 5. Linear sweep voltammetry on a Pt electrode in various ionic liquid media under inert atmosphere at 100 °C, sweep rate = 20 mV·s^{-1}.

The cathodic stability was identical for the three electrolytes, and limited by the reduction of the BMIM$^+$ cation [34] at -1.5 V vs. Ag^{+1}/Ag. We have previously demonstrated that Pt cannot be oxidized in BMIM TFSI [32]. The oxidation signal observed at 2.8 V vs. Ag^{+I}/Ag in BMIM TFSI and in the IL mixture corresponds then to the oxidation of the TFSI$^-$ anion [34] (7):

$$N(SO_2CF_3)_2^- \rightarrow \cdot N(SO_2CF_3)_2 + e^- \quad (7)$$

It has been previously shown that the anodic signal observed in BMIM Cl for potential values was higher than 1 V vs. Ag^{+I}/Ag, and the oxidation peak A_1 recorded in 0.25 M BMIM Cl in BMIM TFSI corresponded to both of the two following oxidation reactions (8) [35,36] and (9), whose occurrence cannot be distinguished from each other in the voltammograms:

$$3\, Cl^- \rightarrow Cl_3^- + 2\, e^- \quad (8)$$

$$Pt + x\, Cl^- \rightarrow PtCl_x^{y-x} + y\, e^- \quad (9)$$

The oxidation of chloride ions must be of little significance for high leaching faradaic yields, and also to avoid appreciable electrolyte degradation.

The speciation of platinum and the efficiency of the leaching step were studied depending on the electrode potential in the range 1.15–2.1 V vs. Ag^{+I}/Ag. Quantitative analysis of platinum content in the electrolyte after leaching was performed by atomic absorption spectrometry (AAS), using analytical procedures specific to ILs media developed previously [32]. Speciation of dissolved platinum was determined by electroanalytical techniques.

Figure 6 compares a voltammogram obtained in the medium after potentiostatic leaching at E = 1.4 V vs. Ag^{+I}/Ag, to that recorded in the same IL mixture containing 15 mM $PtCl_4$: the two voltammograms were similar to whatever potential value was applied for leaching in the above range. Two reduction signals were seen on the i-E curves: peak C_1 at -0.1 V and peak C_2 near -1.3 V vs. Ag^{+I}/Ag. No platinum deposit was obtained by applying a constant potential corresponding to C_1, this signal was therefore attributed to the reduction of Pt^{+IV} to Pt^{+II}. On the contrary, pure platinum could be deposited at -1.3 V vs. Ag^{+I}/Ag, which means that C_2 corresponds to the reduction of Pt^{+II} to Pt^0. It can then be concluded that platinum is leached in the form of Pt^{+IV} beyond 1 V vs. Ag^{+I}/Ag.

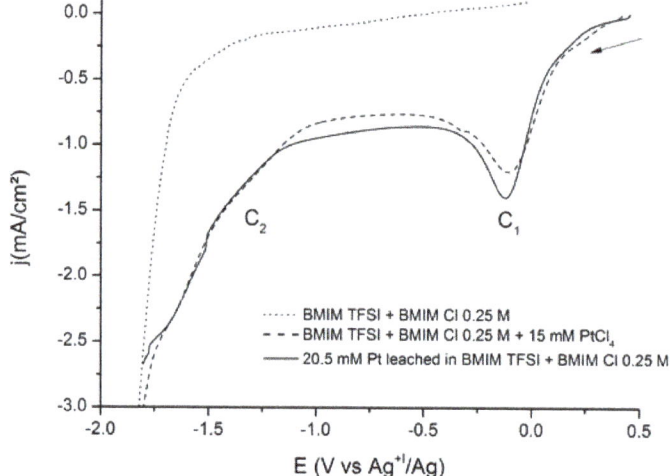

Figure 6. Linear sweep voltammetry performed on a glassy carbon electrode in IL melts under inert atmosphere, $T°$ = 100 °C, sweep rate = 20 mV·s^{-1}.

According to the literature [28,37,38], Pt^{+IV} should be dissolved as a chloride complex. This hypothesis was confirmed by amperometric titration of free chloride ions before and after platinum leaching. It was found that six chloride ions are consumed from the leaching of one Pt atom, which indicates that leached Pt in the IL mixture was in the form of a PtCl$_6^{2-}$ complex, as is usually observed in aqueous media.

Chronoamperometric tests of Pt electroleaching were carried out with 1 cm^2 Pt anode, with a glassy carbon cathode of comparable area. Runs were conducted for a given charge pass, controlling the anode potential. Figure 7 clearly shows that the anode potential had a strong influence on platinum leaching, for both the leaching rate—deduced from the current density—and the faradaic yield. The higher the applied potential, the lower the leaching rate and the faradaic yield: this means that chloride oxidation is favored at higher potential values. Faradaic yield up to 100% can be reached at "low" applied potential, i.e., below 1.3 V vs. Ag^{+I}/Ag, which means the electrolyte degradation by chloride ions oxidation is very little significant in this domain. The corresponding leaching rate is in the order of 3.3 mg·h^{-1}·cm^{-2}, which is comparable to the rate obtained in aqua regia solution reported at 4 mg·h^{-1}·cm^{-2} in [25]. Taking into account the molecular weight of Pt and the leaching involving four electrons, this corresponded to a current density of a few mA·cm^{-2}.

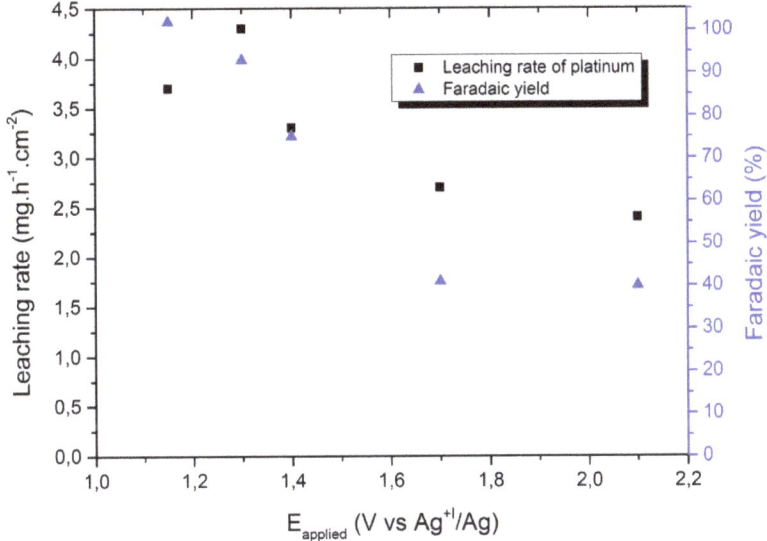

Figure 7. Influence of the applied potential on the leaching rate of platinum and corresponding faradaic yield in runs conducted at Q = 55 C·cm^{-2}.

Electrodeposition of Pt was then performed in Pt-containing IL mixture, with a 1 cm^2 glassy carbon cathode. The cathode potential was controlled in chronoamperometric runs in a range corresponding to the reduction of Pt^{+II} to Pt^0, using a Pt plate as an anode. This latter was electrochemically leached during the electrodeposition of Pt at the cathode. It was preferential to add Pt salts at 7.5 mM to allow the occurrence of cathodic reactions in the first minutes of the runs. Deposition of Pt on the cathode was observed to be effective with a cathode potential at -1.3 V vs. Ag^{+I}/Ag in the 10 h-long test. The anode potential was stable and equal to 0.91 V vs. Ag^{+I}/Ag, a potential value that corresponds to maximum faradaic yield conditions.

Platinum recovery from conducting waste by electroleaching and electrochemical deposition (E/E) in a single cell can then be considered by controlling the cathode potential, without degradation of the electrolyte.

3.2.2. Application to the Recycling of Platinum Contained in PEMFC

This experimental procedure was applied to the recovery of Pt used as catalyst in PEMFC. PEMFC electrodes are carbon-based substrates on which Pt nanoparticles are deposited. A sample of PEMFC electrode with an area of 1 cm^2 was placed as an anode, and the cathode was a glassy carbon plate. Because of the very small amount of Pt—nearly 75 µg·cm^{-2} in the used PEMFC electrode—the IL electrolyte was enriched in Pt prior PEMFC electrode treatment by electrochemical leaching of a platinum plate: as expected by the charge passed, the concentration of Pt (IV) ions produced was measured by AAS at 7.5 mM. The combined electroleaching/electrodeposition test was conducted for 10 h.

The SEM aspect of the fuel cell electrode before (a) and after (b) treatment is shown in Figure 8, on the left side. The black part of the SEM image (a) corresponds to the carbon substrate whereas grey areas are for platinum nanoparticles deposited on the electrode. All the platinum nanoparticles of the PEMFC electrode were leached during the 10 h run (b)—as confirmed by EDX measurements—whereas pure platinum was deposited at the cathode (c). The Pt concentration in the IL bath was not visibly changed by the run.

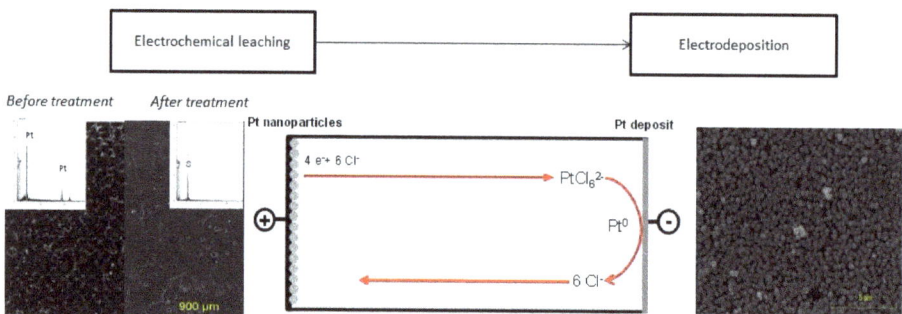

Figure 8. Electrochemical recovery of platinum used in PEMFC electrodes in a 1-butyl-3-methyl imidazolium bis (trifluoromethyl) sulfonyl imide (BMIM TFSI) + BMIM Cl melt containing 7.5 mM of leached Pt under inert and ambient atmosphere.

Coupling of electroleaching to electrodeposition of Pt in IL media has been shown to be a promising route for the recovery of Pt from spent PEMFC. To date, the Pt catalyst represents around 45% of the total cost of a PEMFC so Pt recycling could be a promising route to reduce the production costs. The current, conventional process for Pt recovery from spent catalysts is very complex, involving several steps, some of them requiring organic solvents or strong acidic solutions e.g., aqua regia.

Further works will consist in the study of experimental parameters that influence the electrodeposition faradaic yield and the morphology of the platinum deposit (applied potential/current density, Pt initial concentration of the electrolyte, nature of the substrate, etc.). The global process parameters can then be optimized for the sake of high recovery yield at an acceptable rate, together with a as long a lifetime of the electrolyte as possible.

This work demonstrates once more that ILs are suitable media for the recovery of precious metals. More specifically, IL melts are promising, since physical and chemical properties of the electrolyte can be tuned to the nature of the metals to be recovered, with a high selectivity even from waste containing numerous other elements.

4. Conclusions

Coupling direct or assisted electroleaching to electrodeposition (E/E) in the same cell is an innovative technology classified as a hydrometallurgical hybrid process. This method allows the two key steps of the hydrometallurgical processes to be performed in the same reactor, as shown in several worked examples. E/E could be successfully applied to different kinds of waste: electrode materials from end-of-life spent batteries and accumulators (Zn/MnO$_2$, Ni/Cd), Waelz oxide, and more recently, MEAs issued from used PEM fuel cells. Whatever the anolyte used—aqueous or ILs—leaching can be performed in soft conditions, avoiding the use of concentrated acids. In ILs, precious metals can be oxidized without hazardous ligands such as cyanide ions. Regarding leaching selectivity, in the examples shown with aqueous solutions, leaching relies on electrogenerated protons, and high selectivity would not be obtained in all conditions; in contrast, in ILs with direct electrochemical leaching, applying the suitable anode potential for the targeted element results in high dissolution selectivity. In aqueous media, selective electrodeposition of the targeted metal is feasible, even with complex catholytes, as shown with the (Cd^{2+}, Co^{2+}, Ni^{2+}) catholyte obtained from the black mass issued from spent Ni/Cd accumulators. Use of ILs as an electrolyte allows recovery at the cathode of metals whose reduction cannot be envisaged in aqueous media.

For coupled electro-assisted leaching/electrodeposition in aqueous media, existing published data and developed chemical engineering methodologies are often sufficient to consider in the short term for a pilot process to be designed. In contrast, E/E in ILs is a very new concept for which additional investigations are required for possible development of this promising route; in particular on (i) improved knowledge in metal speciation in ILs with suitable analytic methods; (ii) better durability of the ILs used in a process with nearly perfect selectivity of electrode processes, and (iii) enhancement of current densities for higher production rates by improvement of IL properties and cell designs.

Author Contributions: M.B.: E/E in ILs; C.H.: Ni-Cd E/E treatment in aqueous medium; J.C.: Ni-Cd E/E treatment in aqueous medium; N.L., F.L.: co-supervisors of aqueous medium studies; S.L., E.B.: co-supervisors of ILs studies; E.M.: scientific leader of hydrometallurgical processes research theme and co-supervisor of M.B. and C.H. PhD thesis.

Funding: Part of the work presented here has been completed within the CEATech program with funding from Region Lorraine. Most facilities used have been co-funded within the Contract "Sustainable Chemistry and Processes" (CPER) between French government and Region Lorraine, with contribution of European Union (FEDER).

Conflicts of Interest: The authors declare no conflicts of interest.

References

1. Venkateswaran, K.V.; Ramachandran, P. Electroleaching of sulfides—A review. *Bull. Electrochem.* **1985**, *1*, 147–155.
2. Kirk, D.W.; Gehring, R.; Graydon, W.F. Electrochemical leaching of a silver arsenopyrite ore. *Hydrometallurgy* **1987**, *17*, 155–166. [CrossRef]
3. Elsherief, A.E. A study of the electroleaching of manganese ore. *Hydrometallurgy* **2000**, *55*, 311–326. [CrossRef]
4. Kumari, A.; Natarajan, K.A. Electroleaching of polymetallic ocean nodules to recover copper, nickel and cobalt. *Miner. Eng.* **2001**, *14*, 877–886. [CrossRef]
5. Pozzo, R.L.; Malicsi, A.S.; Iwasaki, I. Removal of lead from printed-circuit board scrap by an electrodissolution-delamination method. *Resour. Conserv. Recycl.* **1991**, *5*, 21–34. [CrossRef]
6. Soare, V.; Dumitrescu, D.; Burada, M.; Constantin, I.; Soare, V.; Capota, P.; Popescu, A.M.; Constantin, V. Recovery of Metals from Waste Electrical and Electronic Equipment (WEEE) by Anodic Dissolution. *REV Chim-Buchar.* **2016**, *67*, 920–924.
7. Kim, E.Y.; Kim, M.S.; Lee, J.C.; Pandey, B.D. Selective recovery of gold from waste mobile phone PCBs by hydrometallurgical process. *J. Hazard. Mater.* **2011**, *198*, 206–215. [CrossRef] [PubMed]
8. Abbott, A.P.; Frisch, G. *Ionometallurgy: Processing of Metals Using Ionic Liquids, in Element Recovery and Sustainability*; Andrew, J.H., Ed.; The Royal Society of Chemistry: Cambridge, UK, 2013; pp. 59–79.
9. Abbott, A.P.; McKenzie, K.J. Application of ionic liquids to the electrodeposition of metals. *Phys. Chem. Chem. Phys.* **2006**, *8*, 4265–4279. [CrossRef] [PubMed]

10. Page, P.W.; Brandon, N.P.; Mahmood, M.N.; Fogarty, P.O. One-step recovery of nickel by an electrohydrometallurgical process. *J. Appl. Electrochem.* **1992**, *22*, 779–786. [CrossRef]
11. Guillaume, P. Recherche d'un Protocole de Traitement de Solides Zincifères Par Voie Électrochimique: Couplage Électrolixiviation/Électrodéposition Dans Une Cellule Unitaire. Ph.D. Thesis, University of Metz, Metz, France, 2008.
12. Guimaraes, Y.F.; Santos, I.D.; Dutra, A.J.B. Direct recovery of copper from printed circuit boards (PCBs) powder concentrate by a simultaneous electroleaching-electrodeposition process. *Hydrometallurgy* **2014**, *149*, 63–70. [CrossRef]
13. Hazotte, C. Traitement de la Matière active D'accumulateurs Ni-Cd en fin de vie par Couplage Electrolixiviation/Electrodéposition. Ph.D. Thesis, University of Lorraine, Lorraine, France, 2014.
14. Balva, M. Etude du Recyclage en Milieu Liquide Ionique de Platinoïdes Immobilisés sur Support Solide Pour une Valorisation Dans la Filière Pile à Combustible. Ph.D. Thesis, University of Lorraine, Lorraine, France, 2017.
15. Urzúa-Abarca, D.A.; Fuentes-Aceitunoa, J.C.; Uribe-Salas, A.; Lee, J.C. An electrochemical study of silver recovery in thiosulfate solutions. A window towards the development of a simultaneous electroleaching-electrodeposition process. *Hydrometallurgy* **2018**, *176*, 104–117. [CrossRef]
16. Guillaume, P.; Leclerc, N.; Lapicque, F.; Boulanger, C. Electroleaching and electrodeposition of zinc in a single-cell process for the treatment of solid waste. *J. Hazard. Mater.* **2008**, *152*, 85–92. [CrossRef] [PubMed]
17. Hazotte, C.; Leclerc, N.; Meux, E.; Lapicque, F. Direct recovery of cadmium and nickel from Ni-Cd spent batteries by electroassisted leaching and electrodeposition in a single-cell process. *Hydrometallurgy* **2016**, *162*, 94–103. [CrossRef]
18. Hazotte, C.; Leclerc, N.; Diliberto, S.; Meux, E.; Lapicque, F. End-of-life nickel–cadmium accumulators: Characterization of electrode materials and industrial Black Mass. *Environ. Technol.* **2015**, *36*, 796–805. [CrossRef] [PubMed]
19. Comel, J. *Optimisation du Traitement D'accumulateurs Ni/Cd par Couplage Électrolixiviation/Électrodéposition*; Internal Report; Institut Jean Lamour, Université de Lorraine: Metz, France, 2017.
20. Hazotte, C.; Meux, E.; Leclerc, N.; Lapicque, F. Electroassisted leaching of black mass solids from Ni-Cd batteries for metal recovery: Investigation of transport and transfer phenomena coupled to reactions. *Chem. Eng. Process.* **2015**, *96*, 83–93. [CrossRef]
21. Guillaume, P.; Leclerc, N.; Boulanger, C.; Lecuire, J.M.; Lapicque, F. Investigation of optimal conditions for zinc electrowinning from aqueous sulfuric acid electrolytes. *J. Appl. Electrochem.* **2007**, *37*, 1237–1243. [CrossRef]
22. Devilliers, D.; Tillemant, O.; Vogler, M. Mise au point sur l'activité et les réalités électrochimiques en France et au plan international. *L'actualité Chim.* **1992**, *1*, 5–34.
23. Silvester, D.S.; Rogers, E.I.; Compton, R.G.; Mckenzie, K.J.; Ryder, K.S.; Endres, F.; Macfarlane, D.; Abbott, A.P. Technical Aspects. In *Electrodeposition from Ionic Liquids*; Wiley-VCH Verlag GmbH & Co. KGaA: Weinheim, Germany, 2008; pp. 287–351.
24. Ohno, H. Physical Properties of Ionic Liquids for Electrochemical Applications. In *Electrodeposition from Ionic Liquids*; Wiley-VCH Verlag GmbH & Co. KGaA: Weinheim, Germany, 2008; pp. 47–82.
25. Li, Q.; Jiang, J.; Li, G.; Zhao, W.; Zhao, X.; Mu, T. The electrochemical stability of ionic liquids and deep eutectic solvents. *Sci. China Chem.* **2016**, *59*, 571–577. [CrossRef]
26. Endres, F. Ionic liquids: Solvents for the electrodeposition of metals and semiconductors. *ChemPhysChem* **2002**, *3*, 144–154. [CrossRef]
27. Billy, E. Application des Liquides Ioniques à la Valorisation des Métaux Précieux Par Une vOie de Chimie Verte. Ph.D. Thesis, University of Grenoble, Saint-Martin-d'Hères, France, 2012.
28. Deferm, C.; Hulsegge, J.; Moller, C.; Thijs, B. Electrochemical dissolution of metallic platinum in ionic liquids. *J. Appl. Electrochem.* **2013**, *43*, 789–796. [CrossRef]
29. Barrosse-Antle, L.E.; Bond, A.M.; Compton, R.G.; O'Mahony, A.M.; Rogers, E.I.; Silvester, D.S. Voltammetry in Room Temperature Ionic Liquids: Comparisons and Contrasts with Conventional Electrochemical Solvents. *Chem. Asian J.* **2010**, *5*, 202–230. [CrossRef] [PubMed]
30. Abbott, A.P.; Frisch, G.; Hartley, J.; Wrya, O.; Karima, W.O.; Rydera, K.S. Anodic dissolution of metals in ionic liquids. *Prog. Nat. Sci. Mater. Int.* **2015**, *25*, 595–602. [CrossRef]

31. Wu, J.; Zhu, X.; Li, H.N.; Su, L.; Yang, K.; Cheng, X.R.; Yang, G.Q.; Liu, J. Combined Raman Scattering and X-ray Diffraction Study of Phase Transition of the Ionic Liquid BMIM TFSI Under High Pressure. *J. Solut. Chem.* **2015**, *44*, 2106–2116. [CrossRef]
32. Balva, M.; Legeai, S.; Leclerc, N.; Meux, E.; Billy, E. Environmentally friendly recycling of fuel cell's membrane electrode assembly using ionic liquids. *ChemSusChem* **2017**, *10*, 2922–2935. [CrossRef] [PubMed]
33. Balva, M.; Legeai, S.; Leclerc, N.; Billy, E.; Meux, E. Procédé de Récupération de Platine, Par Voie Électrochimique, à Partir d'un Matériau Dans Lequel il est Contenu. Patent N EN: 16 56293, 1 July 2016.
34. De Vos, N.; Maton, C.; Stevens, C.V. Electrochemical Stability of Ionic Liquids: General Influences and Degradation Mechanisms. *Chemelectrochem* **2014**, *1*, 1258–1270. [CrossRef]
35. Aldous, L.; Silvester, D.S.; Villagran, C.; Pitner, W.R.; Compton, R.G.; Lagunas, C.; Hardacre, C. Electrochemical studies of gold and chloride in ionic liquids. *New J. Chem.* **2006**, *30*, 1576–1583. [CrossRef]
36. Zhang, Q.B.; Hua, Y.X.; Wang, R. Anodic oxidation of chloride ions in 1-butyl-3-methyl-limidazolium tetrafluoroborate ionic liquid. *Electrochim. Acta* **2013**, *105*, 419–423. [CrossRef]
37. Katayama, Y.; Endo, T.; Miura, T.; Toshima, K. Electrode Reactions of Platinum Bromide Complexes in an Amide-Type Ionic Liquid. *J. Electrochem. Soc.* **2013**, *160*, D423–D427. [CrossRef]
38. Huang, J.F.; Chen, H.Y. Heat-Assisted Electrodissolution of Platinum in an Ionic Liquid. *Angew. Chem. Int. Edit.* **2012**, *51*, 1684–1688. [CrossRef] [PubMed]

© 2018 by the authors. Licensee MDPI, Basel, Switzerland. This article is an open access article distributed under the terms and conditions of the Creative Commons Attribution (CC BY) license (http://creativecommons.org/licenses/by/4.0/).

Article

Novel Task Specific Ionic Liquids to Remove Heavy Metals from Aqueous Effluents

Pape Diaba Diabate [1], Laurent Dupont [1,2,*], Stéphanie Boudesocque [1,2] and Aminou Mohamadou [1,2]

[1] Institut de Chimie Moléculaire de Reims (ICMR), Université de Reims Champagne-Ardenne, CNRS UMR 7312, UFR des Sciences Exactes et Naturelles, Bâtiment 18 Europol'Agro, BP 1039, F-51687 Reims CEDEX 2, France; pape-diaba.diabate@univ-reims.fr (P.D.D.); stephanie.boudesocque@univ-reims.fr (S.B.); aminou.mohamadou@univ-reims.fr (A.M.)
[2] French Network of Hydrometallurgy Promethee, GDR CNRS 3749, 2 Rue du Doyen Roubault, 54518 Vandoeuvre-lès-Nancy CEDEX, France
* Correspondence: laurent.dupont@univ-reims.fr; Tel.: +33-3-2691-3336

Received: 10 May 2018; Accepted: 29 May 2018; Published: 2 June 2018

Abstract: Task Specific Ionic Liquids (ILs) were generated by association between a cationic ester derivative of betaine and coordinating inorganic anions such as dicyanamide (Dca$^-$), chlorosalycilate (ClSal) and saccharinate (sac). Extraction of Cu(II), Ni(II), Co.(II), Pb(II) and Cd(II) from water was performed with these ILs at room temperature. Our results show that ionic liquid with Clsal anions have a high extraction efficiency towards Cu(II), Ni(II), Cd(II), and Pb(II), whereas dicyanamide ionic liquid may extract efficiently Cu(II), Ni(II) Co.(II) and Cd(II). Ionic liquids with saccharinate anions are selective of Cd(II) ions. The extraction mechanism has been studied by the determination of the coextraction of the counter ion of the metal salt. Our results show that the extraction mechanism proceeds via a mixed process involving both cation exchange and ion-pairing. The proportion of which depends on the nature of the cation. The coordination of Cu(II), Ni(II) and Co.(II) in ionic liquid phase was followed by UV-vis spectroscopies. The metal could be back-extracted from the ionic liquid phase with aqueous EDTA solutions. The metal extractability of the ionic liquid after the back-extraction is equivalent to that of the fresh mixture showing that ionic liquid can be reused for several extraction and back-extraction cycles.

Keywords: ionic liquids; metal extraction; liquid-liquid extraction; back-extraction; reusability

1. Introduction

The rapid growth of industrial activities in recent years has led to a significant increase in the volume and toxicity of industrial effluents containing heavy metals. Heavy metals have a significant toxicity towards humans and the environment [1], whence the profusion of regulations governing the treatment and removal of heavy metals from industrial effluents.

Pollution reduction will, in the future, generate economic profits, because of the continuous increase in the value of metals [2]. The elimination of metals from industrial effluents may be achieved by technologies such as chemical precipitation, coagulation, solvent extraction, electrolysis, membrane separation, ion-exchange, and adsorption [3,4]. Among them, liquid-liquid extraction is one of the most performing technologies for the recovery of metal ions, from industrial wastewaters. This technology uses extracting agents and organic solvents (kerosene, toluene, etc.) as diluent. The loss of organic diluent via volatilization, during extraction processing, generates negative environmental impact and may cause serious damage on human health. Consequently, "greener" extraction methods are being sought, and the use of ionic liquids (ILs) constitutes a possible alternative for the replacement of traditional organic solvent [5]. In the last decade, significant works have shown that Room

Temperature Ionic Liquids (RTILs) are potential substitutes for traditional solvents in liquid-liquid extraction processes, for the separation of metal ions [6–9]. Another specific advantage of ILs concerns the possibility of metal recovery by electrodeposition [10]. However, the conventional ionic liquids have a limited efficiency for metal extraction. The use of Task-Specific Ionic Liquids (TSIL) [11–19] by functionalizing organic cation with chelating moieties to increase the affinity of metals for the IL phase may overcome this problem. Ionic liquids with a coordinating anion represent another strategy to increase the recovery of metals by the IL phase. ILs with fluorinated acetylacetonate ligands have shown interesting effectiveness for the extraction of Eu(III), Nd(III), and Co. (II) [20,21]. The development of "green" extraction processes with IL requires a knowledge of the mechanisms involved, for the transfer of ions between IL and the aqueous phase, to privilege the use of ionic liquids limiting the ionic exchanges between the aqueous and organic phases [22–26]. Here, we report the ability of task-specific ionic liquids to remove heavy metals from water. The quaternary ammonium cations derivative of betaine {tri(n-butyl)(-ethoxy-2-oxoalkyl)ammonium (BuNC$_n^+$)} were associated to non-fluorinated coordinating anion such as sacharinate (Sac = $C_7H_4NO_3S^-$), chlorosalicylate (ClSal = $C_7H_5O_3^-$) and dicyanamide (Dca = $C_2N_3^-$) to generate hydrophobic ionic liquids (Figure 1), use as pure extracting phase [13,27,28]. The potential of ILs with chelating anions, as extracting agents, was investigated towards divalent toxic metals Cu(II), Cd(II), Ni(II), Co.(II) and Pb(II). The choice of the betaine derivative is justified by its accessibility via simple synthetic route, by its availability, the cost of starting materials, and its structural modularity, which allows the control of the hydrophobicity of cation by varying the alkyl chain length bound to the ammonium group. The choice of anions is dictated by their hydrophobic and chelating nature. The sacharinate and chlorosalicylate anions [29,30] are known for their complexing capacity towards heavy and first row transition metals, respectively. The dicyanamide anion is a cheap anion that is easy to handle to generate ILs. Its chelating ability for metal ions is well known [31,32] and the ability of the Dca$^-$ ionic liquids to extract Cu(II) and Ni(II) from water [26,33,34] was shown in a previous study.

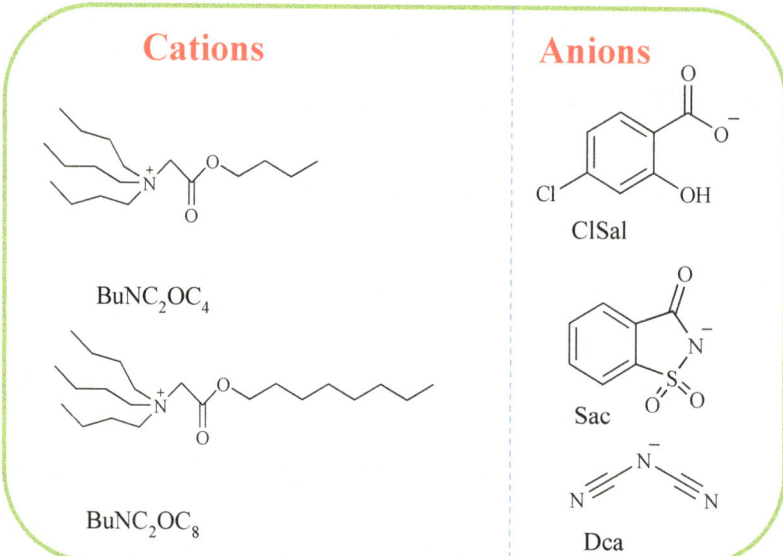

Figure 1. Structure of different analogues of glycine betaine based ionic liquids used in this study.

2. Experimental Section

2.1. Chemicals and Reagents

All chemical and reagents used in this study were used as received without further purification. Sodium nitrate (99%) sodium chloride (99%), sodium dicyanamide (99%), Tri(n-butyl)amine (99%), 1-butanol (99%) and ethyl bromoacetate (98%) were obtained from Sigma-Aldrich (Diegem, Belgium). Ethylenediaminetetraacetic acid disodium salt dihydrate (99%), methanesulfonic acid (99%), 1-octanol (99%), sodium saccharinate (99%) and 4-chlorosalicylic acid (99%) was purchased by Acros (Illkirsh, France).

The solutions of metals were prepared by dissolving their corresponding nitrate salt (analytical grade purchased from Sigma-Aldrich (Saint Quentin-Fallavier, France)-Fluka Chemical (Bucharest, Romania,) in double distilled and deionized water.

2.2. Analytical Measurements

Elemental analyses (C, H, and N) were carried out on a Perkin-Elmer 2400 C, H, N element analyzer in our university. The UV-visible spectra of metal solutions were recorded using a Carry-5000 Varian spectrophotometer. Routine ^{1}H NMR spectra were recorded in deuterated dimethyl sulfoxide C_2D_6OS at room temperature with a Bruker AC 250 spectrometer. Chemical shifts (in ppm) for ^{1}H NMR spectra were referenced to residual protic solvent peaks. The metal analyses were performed by UV-vis spectroscopy using EDTA for Cu(II) and Ni(II) and by ICP-OES, Thermo Fisher ICAP Series for Cd(II) and Pb(II). The concentrations of NO_3^- ions in aqueous solution before and after extraction were determined by ion-HPLC with a Metrohm with a conductivity detector.

2.3. Extraction Experiments

Metals nitrate aqueous solutions were prepared in deionized water. Metal ions distribution ratios were determined by mixing 0.5 g of IL and 2 mL of aqueous phase. The mixture was shaken for 24 h to reach equilibrium and then centrifuged at 2000 rpm for 5 min. The separated organic and aqueous phases, both clear and transparent, were then separated for analysis. The aqueous phase composition was analyzed by spectrophotometry UV-vis or by ICP-OES. The metal ion concentration in the IL phase was deduced from the difference between the concentration of metal ions in the aqueous phase before and after extraction. The efficiency of the extraction process was evaluated by calculation of the extraction percentage (% E) using the following equation:

$$E = 100 * \frac{(C_{in} - C_{fin})}{C_{in}} \quad (1)$$

where C_{in} (mol L^{-1}) is the concentration in the initial aqueous solution and C_{fin} (mol L^{-1}) is the concentration in the final aqueous solution. The metal extraction percentages (% E) were determined at 25 °C. The initial concentration of metal solutions is fixed at 5×10^{-2} mol L^{-1}. The experiments were made in triplicate to ensure the reproducibility of the assay, and the mean values of extraction yields were considered for each system studied. The distribution ratio (D) is calculated using the following formula:

$$D = \frac{(C_{in} - C_{fin})}{C_{fin}} \frac{V_w}{V_{IL}} \quad (2)$$

V_w and V_{IL} correspond to the volume of water and ionic liquid phases, respectively. The maximum D value measurable in this study is assumed to be 5×10^2. The relative uncertainty on D is ±10%. Experimental results done in duplicate agree within 5%.

In back-extraction experiments, the IL phase with metal extracted was contacted with 2 mL of aqueous disodique EDTA solution (C = 10^{-1} mol·L^{-1} of Na_2H_2Y) during 4 h under stirring. The percentage of metal back extracted has been determined from the analysis of aqueous phase.

3. Results and Discussion

The preparation of different ILs was carried out by three main steps: the first step is the preparation of bromide precursor ($BuNC_2OC_2$-Br) and the second step is the preparation of the cationic ester derivative of betaine by transesterification of $BuNC_2OC_2$-Br with butanol or octanol. The third step is the anionic metathesis reaction to generate hydrophobic ILs. The synthetic pathway is given in Figure 2.

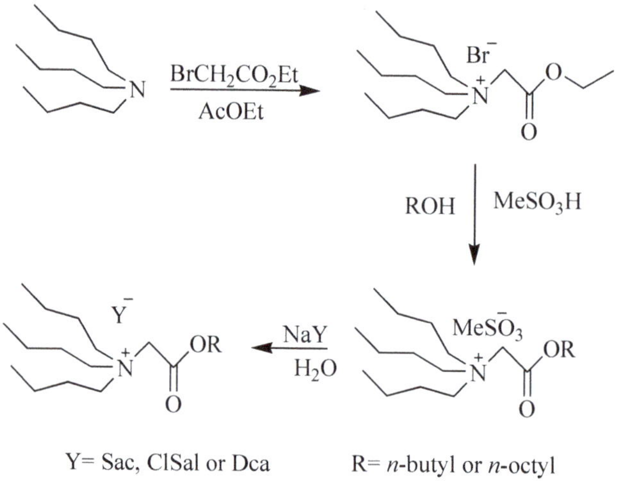

Figure 2. Synthetic route of hydrophobic ionic liquids.

The preparation of bromide salt can be considered as the least difficult step; this reaction is performed in mild conditions. The bromide salt powder was recovered through simple filtration with a quantitatively yield.

In the second step, the strategy used for the esterification reaction is to perform the reaction without any additional solvent; alcohol was also considered as solvent. Methanesulfonic acid was selected as reaction catalyst. It is the best candidate for a green synthesis route, often recyclable and less aggressive than conventional acids [35]. Anionic metathesis between the ester formed and sodium saccharinate, sodium dicyanamide and sodium chlorosalicylate is carried out in water and led in all cases to the formation of hydrophobic ILs with a good yield (>78%). All the ILs are viscous liquids at room temperature.

ILs with saccharinate and chlorosalicylate anions are slightly denser than water with values ranging from 1.040 to 1.111 g·mL^{-1} at 25 °C (Table 1), whereas ILs with dicyanamide anion have the density less than unity. All the ionic liquids with $BuNC_2OC_4^+$ cations are denser than those of $BuNC_2OC_8^+$ cations.

At room temperature, all ionic liquids form two liquid phases when contacted with water. In biphasic system, water is the upper phase with sacharinate and chlorosalicylate based ionic liquids, and the lower phase with dicyanamide ionic liquids.

Table 1. Room temperature ionic liquids density at 25 °C.

Ionic Liquid	Density (g·mL^{-1})
BuNC$_2$OC$_4$-Sac	1.11
BuNC$_2$OC$_8$-Sac	1.04
BuNC$_2$OC$_4$-ClSal	1.10
BuNC$_2$OC$_8$-ClSal	1.02
BuNC$_2$OC$_4$-Dca	0.96
BuNC$_2$OC$_8$-Dca	0.96

The solubility of the ionic liquids in water were measured by NMR for dicyanamide based ionic liquids, or spectrophotometry UV-vis for chlorosalicylate or saccharinate based ionic liquid. The solubility of BuNC$_2$OC$_4$-Clsal and BuNC$_2$OC$_8$-Clsal are equal to 0.3 and 0.005%, respectively and are comparable with those of BuNC$_2$OC$_4$-Dca and BuNC$_2$OC$_8$-Dca equal to 0.35 and 0.26%, respectively. Ionic liquids with saccharinate anion are more soluble than those with dicyanamide and chlorosalycilate anions with the percentage values of 2.85 (BuNC$_2$OC$_4$-Sac) and 1.24% (BuNC$_2$OC$_8$-Sac), respectively.

3.1. Extraction of Cu(II), Ni(II), Cd(II), Co.(II) and Pb(II) from Aqueous Solutions

We compare the extraction properties of six ionic liquids with a complexing anion, towards a panel of five metal cations, Cu(II), Ni(II), Cd(II), Co.(II) and Pb(II). These cations are chosen for their presence in industrial discharges.

The percentage of extraction (%E) is determined with solutions of the nitrate salts of each metal at 0.05 mol·L^{-1} and at 25 °C. The extraction yields (%E) for each IL are depicted in Figure 3. The corresponding distribution ratio are given in Table 2.

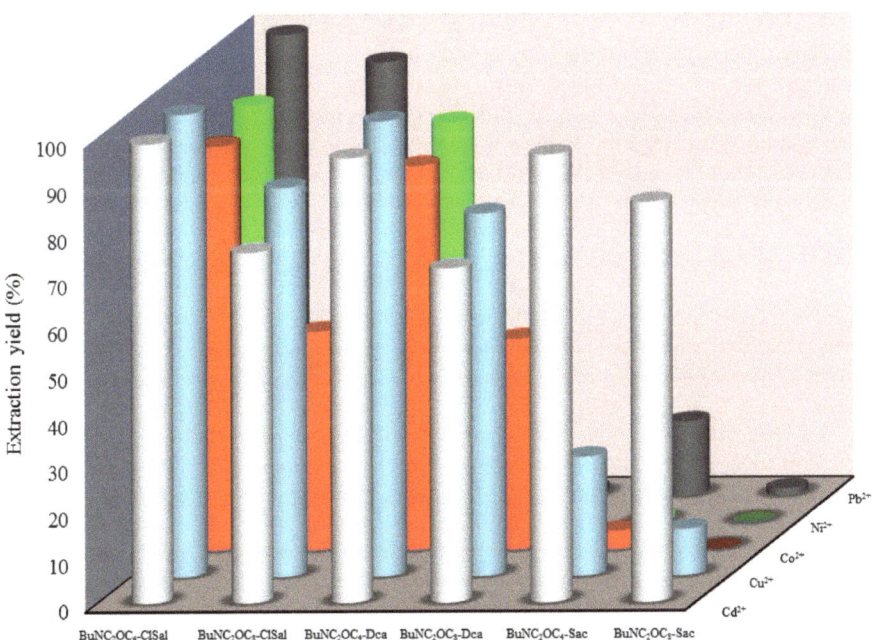

Figure 3. Extraction yields (%E) for metal aqueous nitrate salt with BuNC$_2$OC$_4$ (C8)-Sac, BuNC$_2$OC$_4$(C8)-Clsal and BuNC$_2$OC$_4$(C8)-Dca. C$_{metal}$ = 0.05 mol L^{-1}; V$_w$ = 2 mL; m$_{IL}$ = 0.5 g.

3.2. Influence of the Anion of the Ionic Liquid

The overall examination of Figure 3 shows that the most efficient anion for the extraction of metals is the ClSal anion. Indeed, for all metallic cations, extraction yields are higher than 75% with $BuNC_2OC_4$-ClSal. By comparison, $BuNC_2OC_4$-Dca shows slightly weaker extraction yield, and $BuNC_2OC_4$-sac shows interesting extraction properties only towards Cd(II). The influence of the anion is more pronounced with ionic liquids containing $BuNC_2OC_8$ cation. Indeed, $BuNC_2OC_8$-ClSal and $BuNC_2OC_8$-Dca exhibit similar extraction yield with Cu(II), Cd(II) and Co.(II), but the extraction of Ni(II) does not exceed a few percent with $BuNC_2OC_8$-Dca, while it is close to 50% with $BuNC_2OC_8$-ClSal. Moreover, Pb(II) is totally extracted with $BuNC_2OC_8$-ClSal, whereas the extraction does not exceed a few percent with $BuNC_2OC_8$-Dca.

Table 2. Distribution ratio (D) for metal aqueous nitrate salt with $BuNC_2OC_4(C_8)$-Sac, $BuNC_2OC_4(C_8)$-ClSal and $BuNC_2OC_4(C_8)$-Dca; C_{metal} = 0.05 mol L^{-1}; V_w = 2 mL; m_{Il} = 0.5 g.

Ionic Liquid	Cu(II)	Cd(II)	Pb(II)	Ni(II)	Co.(II)
$BuNC_2OC_4$-Sac	1.5	120.0	0.8	0.1	0.2
$BuNC_2OC_8$-Sac	0.4	25.4	0.1	-	-
$BuNC_2OC_4$-ClSal	1301.4	439.6	735.6	40.0	29.7
$BuNC_2OC_8$-ClSal	21.30	12.8	59.9	4.5	3.7
$BuNC_2OC_4$-Dca	192.5	178.5	0.1	24.1	18.1
$BuNC_2OC_8$-Dca	13.7	1.8	0.1	0.3	3.2

The results reported in Figure 3 and in Table 2 show that the capability of ionic liquids to extract a metal ion is greatly correlated to the ability of the anion of the ionic liquid, to form stable complexes with the metal cation. The extraction of first row transition metal (Cu(II), Co.(II), Ni(II)) by Dca or ClSal based ionic liquids show that the affinity of the metal ions for ionic liquid phase is weak for Ni(II) and Co.(II) and higher for Cu(II). Such a trend may be related to the formation constants of complexes of these metals with the anions of the ionic liquids.

This assumption is corroborated by the values of the stability constants of M(II)-salicylate complexes for Cu(II), Ni(II) and Co.(II). The salicylate moieties may form stable relatively complexes by bidentate coordination giving a five membered rings species with metal ions. The first stability constant of M(II)-salycilate complexes (logβ_1) for Ni(II) and Co.(II) are 8.65 and 8.09, respectively. This reflects a similar affinity of these two metals for the salicylate anion [31,36]. The higher extraction yields of Cu(II) compared to Ni(II) and Co.(II) is in agreement with a higher stability constant for Cu(II) with a logβ_1 of 10.65 [31]. The high extraction yields of heavy metals (Cd(II) and Pb(II)) by ClSal based ionic liquids are equally in relation with the high stability constants of salicylate-Cd(II) (or Pb(II)) complexes [36,37]. The selectivity factor (SF) which represents the ratio between the distribution ratio of two metals for a given ionic liquid, indicates the selectivity of the ionic liquid towards the two metals gives useful information about the efficiency of the separation. The selectivity factors of Cu(II) towards Ni(II) (SF$_{Cu/Ni}$) or Co.(II) (SF$_{Cu/Co.}$) for $BuNC_2OC_4$-ClSal are between 30 and 50, respectively, whereas for $BuNC_2OC_8$-ClSal it is only between four and six, respectively. It seems that $BuNC_2OC_4$-ClSal is more selective than $BuNC_2OC_8$-ClSal for the separation of the first row transition metals.

The coordination of metal in the IL phases was investigated on the basis of the spectrophotometric analysis of the ionic liquid phase after extraction of Co.(II), Cu(II) and Ni(II). The electronic spectra of Ni(II) in $BuNC_2OC_8$-ClSal are characteristic to Ni(II) in an octahedral environment (Figure 4). The two bands with the λ_{max} located at 1145 and 700 nm are assigned to the $^3A_{2g} \rightarrow {}^3T_{2g}$ and $^3A_{2g} \rightarrow {}^3T_{1g}$ (F) transitions, respectively [38]. A third band with the λ_{max} located around 400 nm and partially masked by an intense transfer charge band is attributed to a $^3A_{2g} \rightarrow {}^3T_{1g}$ (P) transitions. The spectrum is shifted to weaker energy by comparison to the spectrum of $Ni(H_2O)_6^{2+}$ in aqueous solutions. This evidences the coordination of Ni(II) by oxygen atoms of the ClSal anion.

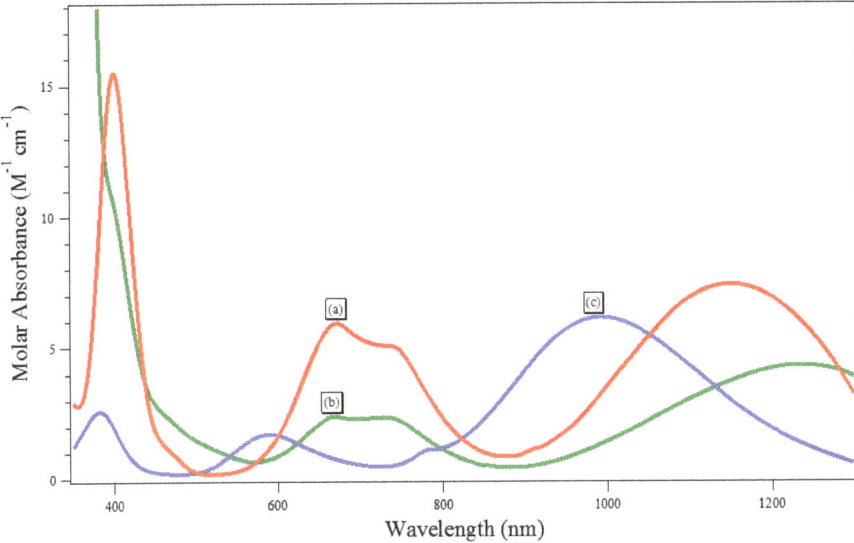

Figure 4. UV-Vis spectra of Ni(NO$_3$)$_2$ in ILs (**a**) BuNC$_2$OC$_4$-ClSal phase; (**b**) BuNC$_2$OC$_4$-Dca and (**c**) aqueous phase. C_{metal} = 0.05 mol L^{-1}; V_w = 2 mL; m_{IL} = 0.5 g; UV-vis spectra are recorded in ethylacetate for ionic liquid (IL) phase.

UV–visible spectra of Cu(II) in IL phases are depicted Figure 5. The spectrum of BuNC$_2$OC$_8$-ClSal after extraction of copper shows one single d–d transition in the visible region with a maximum wavelength at 745 nm. The relatively broad shape of the band and the value of molar extinction coefficient would be indicative of an octahedral copper complex. The higher molar absorbance of Cu(II) in IL phase indicates that the Cu(II) complex is more distorted than the Cu(H$_2$O)$_6^{2+}$ species.

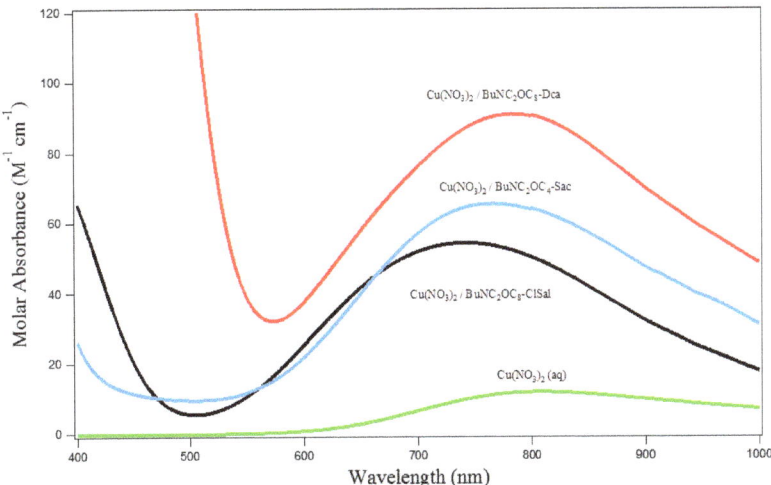

Figure 5. UV-Vis spectra of Cu(NO$_3$)$_2$ in aqueous and in ionic liquid phase. C_{metal} = 0.05 mol L^{-1}; V_w = 2 mL; m_{IL} = 0.5 g; UV-vis spectra are recorded in ethylacetate for IL phase.

Figure 6 represents the spectra of Co^{2+} in the different IL phases. The spectrum of Co.(II) in BuNC$_2$OC$_8$-ClSal shows in the visible region, similar features than the spectrum of Co.(NO$_3$)$_2$ in aqueous phase, with a band at 530 nm and with a shoulder at 480 nm. This band as well as its shoulder correspond to the two following transitions $^4T_1g(F) \to {}^4A_2g(F)$ and $^4T_1g(F) \to {}^4T_1g(P)$. The molar absorption coefficient of IL phase at 530 nm is equal to 33 cm mol^{-1} l. All these features are indicative of a Co.(II) in octahedral environment. It is interesting to note that the spectrum of Co.(II)) in BuNC$_2$OC$_8$-ClSal is shifted towards lower energy compared to those recorded in aqueous media.

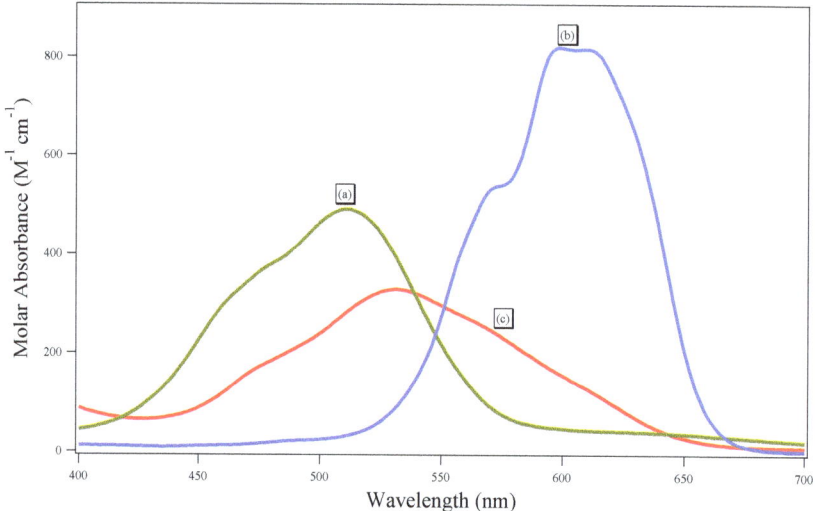

Figure 6. UV-Vis spectra of Co.(NO$_3$)$_2$ in (**a**) BuNC$_2$OC$_8$-Dca phase; (**b**) aqueous phase step up 100 times and; (**c**) BuNC$_2$OC$_8$-ClSal phase step up 10 times. C$_{metal}$ = 0.05 mol L^{-1}; V$_w$ = 2 mL; m$_{Il}$ = 0.5 g; UV-vis spectra are recorded in ethylacetate for IL phase

In previous works, we pointed out the properties of tri(n-butyl)[2-ethoxy-2-oxoethyl]ammonium (BuNC$_2$OC$_2^+$) dicyanamide (Dca) and bis(trifluoromethylsulfonyl)imide (Tf$_2$N$^-$) ionic liquids with for the extraction of metal cations [26,29,30]. Tf$_2$N$^-$ is a non-coordinating anion currently used to generate ionic liquid. BuNC$_2$OC$_2$-Dca extracts quantitatively Cu(II), Ni(II), Cd(II) and Pb(II) from 0.05 mol L^{-1} aqueous solutions. By comparison, for the same metal concentration, BuNC$_2$OC$_2$-Tf$_2$N provides only weak extraction with extraction yield less than 10% [30]. Similar observations have been made with ionic liquid based tetraalkylammonium cation associated with Dca$^-$ and Tf$_2$N$^-$ anions [29]. Although it has high extraction yields, BuNC$_2$OC$_2$-Dca does not allow considering applications in the field of the purification of industrial effluents because of its solubility in water (≈6%), that is why we focused on the development of more hydrophobic ionic liquids, to reduce the releasing of organic cations in aqueous phase. The high extraction properties of Dca$^-$ based ionic liquids are ascribed to the ability of dicyanamide anion to interact with metallic cations through the formation of metal complexes via their nitrogen atoms [33,34]. The examination of the structure of metal complexes in which the dicyanamide anion acts as a ligand, shows that this anion behaves as a monodentate or bridging ligand [33,34]. It is reasonable to think that the higher extraction yields observed for Cu(II) and Cd(II) compared to those of Ni(II), Co.(II) and Pb(II) are related to their higher affinity for Dca$^-$ anion, which favors their transfer from aqueous to ionic liquid phases. However, the lack of thermodynamic data from the literature does not allow corroborating this affirmation. The only published data show that the affinity of the dicyanamide ligand for Cd(II) ions is higher than for Pb(II) with a logβ_1 values of 3.11 for Cd (II) [39] against 2.1 for Pb(II) [40]. These values also show that the affinity of the dicyanamide ligand for the metal cations is lower than in the case of the chlorosalicylate ligand, which is consistent with the lower extraction yields observed with dicyanamide based ionic liquids compared to the chlorosalicylate one. The dicyanamide based ionic liquids are equally less selective for the first row transition metals than their analogous with ClSal anions. The selectivity factors SF(Cu/Ni) and SF(Cu/Co.) are equal to 8 and 10.6 for $BuNC_2OC_4$-Dca and with $BuNC_2OC_8$-Dca, SF(Cu/Co.) is equal to 4.29. The main exception is the relatively high selectivity of Cu(II) towards Ni(II) with $BuNC_2OC_8$-Dca, compared to other systems, with a selectivity factor of 53 due to a low percentage of Ni(II) extracted. It is to be noticed the low affinity of dicyanamide based ionic liquids for Pb(II) ions which makes it possible to use them effectively to separate metals from Pb(II) ions. With $BuNC_2OC_8$-Dca, the selectivity factors SF(Cu/Pb) and SF(Cd/Pb) are equal to 230 and 170.

The UV spectrum of Ni(II) in $BuNC_2OC_8$-Dca exhibit the same features as the spectrum of $Ni(H_2O)_6^{2+}$ in aqueous solution, and is characteristic to Ni(II) in an octahedral environment. The shift towards weaker energies indicates a change in the nature of donor atoms bound to the metal center related to the implication of nitrogen atoms in the coordination sphere of the metal [38].

The UV-visible spectrum of Cu(II) in $BuNC_2OC_8$-Dca show the same features as in $BuNC_2OC_8$-ClSal in the visible region, except that the shift of the d-d transition is less marked than with $BuNC_2OC_8$-ClSal with a maximum at 780 nm. In the UV region, the spectrum shows a supplementary transition centered at 390 nm (not shown here). This transition is ascribed to an LMCT (or MLCT) transition characteristic of the Dca coordination to Cu(II) cation. The spectrum of Co.(II) in $BuNC_2OC_4$ (C_8)-dca shows in the visible region a band at 600 nm with a shoulder at 570 nm. The molar absorption coefficient at 600 nm is equal to 810 cm mol^{-1} l. The spectral features of Co(II) spectra in Dca ionic liquids are characteristic of a Co(II) in tetrahedral environment, meaning a change of coordination of the metal during the extraction process. The band is assigned to a $A_2(F) \rightarrow T_1(P)$ transition [38].

The saccharinate anion behaves as monodentate ligand through its nitrogen atom or as a bidentate through N,O coordination [41]. It often act as a ternary ligand in structure of many metal complexes to achieve the coordination of metal ions [42]. No data are available in the literature concerning the formation constants in aqueous solution of divalent metal complexes with this anion, suggesting a weak coordinating ability in aqueous phase towards metal ions. One key point is that saccharin ionic liquids have a high affinity only for Cd(II), and may also be efficient to separate cadmium from other divalent cations. Indeed, the lowest selectivity factors for cadmium towards other divalent cations ($SF_{(Cd/M)} = D(Cd(II))/D(M(II))$) is that with Cu(II) ions. They are equal to 63 and 80 for $BuNC_2OC_8$-Sac and $BuNC_2OC_4$-Sac, respectively.

3.3. Influence of the Cation of the Ionic Liquid

Figure 3 shows that the cation of the ionic liquid has a significant influence on the extraction of metal cation. Indeed, the extraction yield is significantly higher with $BuNC_2OC_4^+$ than with $BuNC_2OC_8^+$ cations. This trend is related to (i) the more hydrophilic character of the tetrahexylammonium cation which may favor the transfer of the metal in the ionic liquid phase via cationic exchange process, and (ii) the fact that the concentration of anion, in the ionic liquid phase, is higher in $BuNC_2OC_4^+$ ILs than with $BuNC_2OC_8^+$ ILs. Indeed, as an example taking into account the molar mass of $BuNC_2OC_4$-Dca and $BuNC_2OC_8$-Dca, the mole number contacted with the aqueous phase in an extraction sample is respectively of 1.3 mmol for $BuNC_2OC_4$-Dca and 1.1 mmol for $BuNC_2OC_8$-Dca, respectively. Such a difference may be taken into account to explain partially the better extraction properties of $BuNC_2OC_4$-Dca, which contains a higher concentration of extractant. To investigate if ion exchange occurs during extraction, coextraction of nitrate ion was followed by ion-chromatography. When a cation exchange occurs, the metal ion in the aqueous phase is exchanged with a cation of the ionic liquid. The metal ion is then extracted into the ionic liquid, while the cation of the IL moves to the aqueous phase following the equilibria:

$$M^{2+}{}_{(w)} + 2\ BuNC_2OC_n{}^+{}_{(IL)} + nA^-{}_{(IL)} \rightleftharpoons M(A)_n{}^{(n-2)-}{}_{(IL)} + 2\ BuNC_2OC_n{}^+{}_{(w)} \qquad (3)$$

If extraction proceeds only via ion-pairing, the counter anion of the metal salt is co-extracted into the ionic liquid phase, so that the ratio between the metal and the counter-ion extracted should be equal to one considering the following equilibria:

$$M^{2+}{}_{(w)} + 2X^-{}_{(w)} + nA^-{}_{(IL)} \rightleftharpoons M(A)_n{}^{(n-2)-}{}_{(IL)} + 2X^-{}_{(IL)} \qquad (4)$$

The determination of the co-extraction rates of nitrate ion gives an indirect insight of the amount of cation released in the aqueous phase and allow to investigate the extraction mechanism of metal cation. Figure 7 depicts the ratio R between the extraction rate of nitrate ion and those of metal.

Ion-pairing is the dominant mechanism with $BuNC_2OC_8{}^+$ cation. Indeed, for all systems, except for Cu(II)/$BuNC_2OC_8$-ClSal, the ratio $E(A^-)/E(C^+)$ values are close and even may exceed 1 (Figure 7). A value higher than 1 means that the anions of the metal salt are more efficiently extracted than metal itself and suggests the possibility of anion exchange between the anion of the ionic liquid and the nitrate ions. This may take place during the extraction of Co.(II) in $BuNC_2OC_8$-Dca (R = 1.3) and to a small extent with $BuNC_2OC_8$-ClSal (R = 1.1).

The R value is much less than one with $BuNC_2OC_4{}^+$ cation, meaning that the mechanism for extracting metal ions is a mixed process that includes both cation exchange and ion-pair extraction. The R value represents the extent of ion-pair extraction. It is clearly seeing from Figure 7 that with $BuNC_2OC_4$-Clsal, the extraction of metal proceeds mostly via cation-exchange, especially, in the case of the extraction of Ni(II). In the case of $BuNC_2OC_4$-Dca, the two mechanisms are operant, nearly to the same extent.

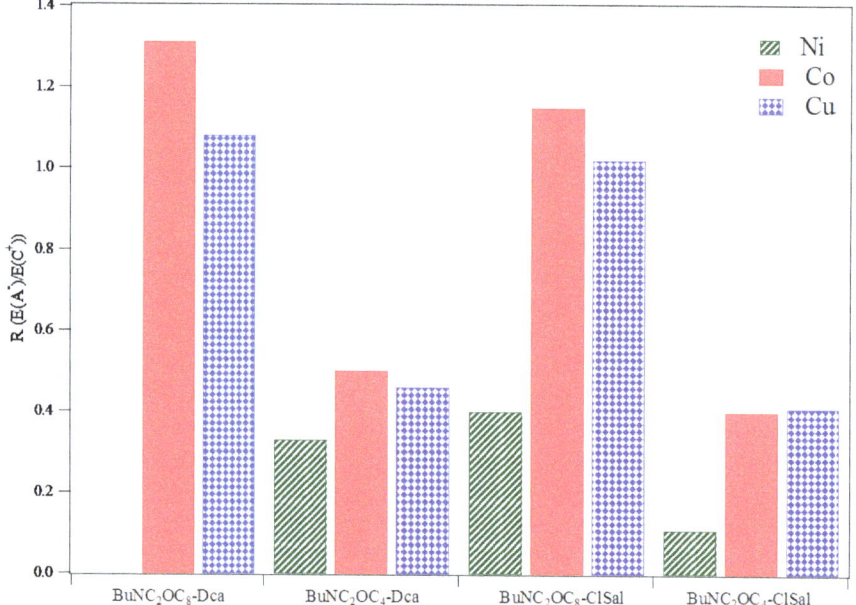

Figure 7. Ratio (A^-/C^+) between the extraction rate of anion and those of Cu(II), Co.(II) and Ni(II), for the different salt used at different concentration.

All the results show that $BuNC_2OC_4^+$ based ionic liquids are more efficient to extract metal ions than that of $BuNC_2OC_8^+$. The disadvantage is that their higher solubility leads to the extraction with a large proportion of cationic exchange resulting in a significant release of organic cations in the aqueous phase. In comparison, $BuNC_2OC_8^+$ based ionic liquids, despite their lower performance, favor exclusively ion-pair extraction and allow the development of more eco-compatible extraction processes. These results may suggest that the $BuNC_2OC_4^+$ based ionic liquids would be less viable in large-scale use. However, it is also necessary to conceive a possible use in media with high salinities, which corresponds to most liquid industrial effluents. As we have shown in previous work [26], high salinity causes an increase in extraction efficiency, promotes ion pair extraction and thus limits the release of organic cations during extraction processes.

3.4. Back-Extraction of Metal Ions and Recyclability of Ionic Liquid

The back-extraction process will be based on exchange ligand reaction and would be suitable by using a hydrophilic ligand with stronger affinity for metal cations than the anions of ionic liquid. To verify this hypothesis, we carried out the experience of back-extraction with aqueous disodique EDTA solution at 0.1 mol L^{-1} contacted with ionic liquid phases previously loaded with metal. EDTA is a hexavalent ligand that forms stable 1:1 complexes with divalent metals; the stability constants of the complexes formed are often higher than 12 (in logarithmic units). In this case, the back-extraction process would involve a replacement of M(II) by Na$^+$ ions in the ionic liquid phase, or the back-extraction of NO$_3^-$ ions previously extracted in the ionic liquid phase, or both of them.

Figure 8 shows that EDTA is an efficient ligand to recover metal ions from the ionic liquid phase. Cu(II) is back extracted from $BuNC_2OC_n$-Clsal with recovery percentages close to 100% and higher than 80% with $BuNC_2OC_n$-Dca. The efficiency is observed for the recovery of Ni(II) Cd(II) and Co.(II) from these two ionic liquids. Cd(II) is also fully back extracted from $BuNC_2OC_n$-Sac with a back extraction percentage exceeding 90%.

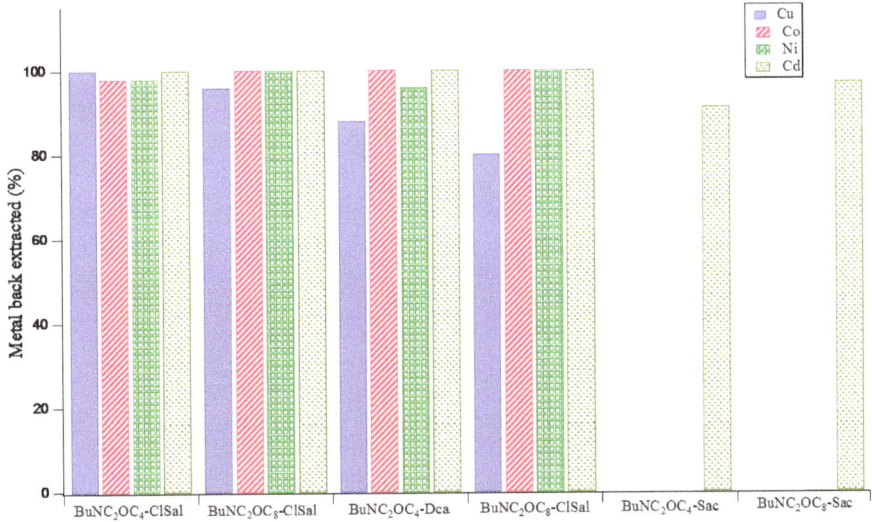

Figure 8. Extraction and back-extraction yield (%E) for M(II).

After the back-extraction process, it is important to check if the ionic liquid would be able to be reused for the extraction of metal from aqueous solution. To check the recyclability of the ionic liquids, several cycles of extraction-back extraction were carried-out on the same ionic liquid phase, following the same experimental conditions as those reported previously.

After each cycle, the metal concentration was determined in the aqueous phase, and the extraction and back-extraction yields have been determined by taking into account the total concentration of metal present in the ionic liquid phase. Figure 9 reports the extraction and back-extraction yields for Cu(II) and Co.(II) with $BuNC_2OC_n$-Sal ionic liquid.

The results show that the efficiency of extraction and back-extraction remained similar over three cycles. The extraction yields are nearly constant over the three cycles, whereas the back-extraction yields are higher than 95%. Therefore, these results confirm that the ionic liquid can be regenerated in the back-extraction process with EDTA solution and can be reused for metal extraction; the release of ionic liquid remains sufficiently limited to not alter the performance of the latter. These results may be generalized for all the systems studied in this work.

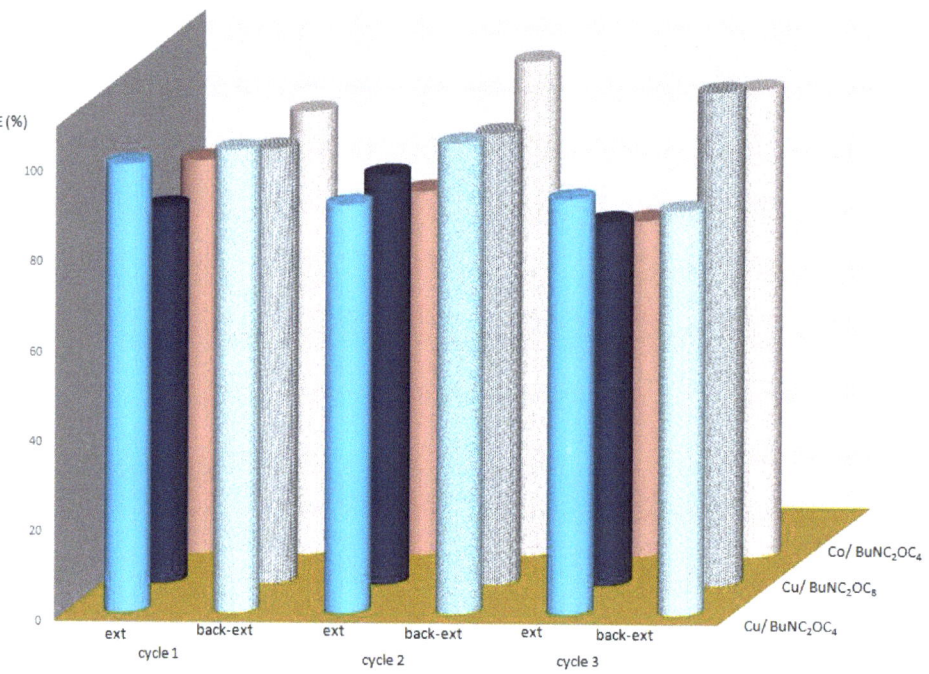

Figure 9. Reusability of ionic liquid phase for copper and cobalt extraction using $BuNC_2OC_n$-Clsal.

4. Conclusions

Our results report the possibility of designing in a simple way hydrophobic ILs able to extract efficiently metallic cations by liquid-liquid extraction. The ionic liquids use as negative pole common coordinating anions, such as dicyanamide, chlrorosalicylate and saccharinate, instead of fluorinated compounds. Our first results show that chlorosalicylate based ionic liquids have a high extraction efficiency towards Cu(II), Ni(II), Cd(II), and Pb(II), whereas dicyanmide based ionic liquids may extract efficiently Cu(II) and Cd(II). Ionic liquids with saccharinate anions are selective of Cd(II) ions. The UV-vis investigations of the first row transition metals show the coordination of the metal with the anion of ionic liquids.

The extraction process is reversible with all the ionic liquids, and the metal ions can be recovered by aqueous solutions of disodique EDTA. In order to conceive environmentally-friendly systems for industrial applications, a workable way is to privilege the ionic liquids with the BuNC$_2$OC$_8$$^+$ cation, which are less soluble. The ionic liquids with the BuNC$_2$OC$_4$$^+$ cation nevertheless remain interesting for the development of liquid-liquid extraction processes, but their extraction properties must be studied in higher salinity environments which favor extraction by ion pair. These results show that the strategy of designing ionic liquids with coordinating anions with low hapticity giving moderately stable complexes in the ionic liquid phase is a promising route for the extraction of aqueous metal ions. In this case, the metals can be easily recovered in the aqueous phase with water-soluble ligands with high hapticity.

Author Contributions: L.D. and S.B. designed the detailed experimental plan for this research. P.D.D. carried out the experiments and data analysis under the supervision of S.B and A.M., L.D. and A.M. wrote the manuscript with help of other coauthors.

Acknowledgments: The authors are grateful to the "Region Grand Est" and the FEDER for a Grant to Pape Diaba Diabate and for its financial support.

Conflicts of Interest: The authors declare no conflict of interest.

References

1. Lichtfouse, E.; Schwarzbauer, J.; Rober, T.D. *Environmental Chemistry: Green Chemistry and Pollutants in Ecosystems*; Springer: Berlin, Germany, 2005; p. 780.
2. Kumbasar, R.A. Extraction and concentration study of cadmium from zinc plant leach solutions by emulsion liquid membrane using trioctylamine as extractant. *Hydrometallurgy* **2009**, *95*, 290–296. [CrossRef]
3. Blais, J.F.; Dufresne, S.; Mercier, G. State of the art of technologies for metal removal from industrial effluents. *Revues Des. Sci. De L'eau* **1999**, *12*, 687–711. [CrossRef]
4. Fu, F.; Wang, Q. Removal of heavy metal ions from wastewaters: A review. *J. Environ. Manag.* **2011**, *92*, 407–418. [CrossRef] [PubMed]
5. Huddleston, G.J.; Willauer, H.D.; Swatloski, R.P.; Visser, A.E.; Rogers, R.D. Room temperature ionic liquids as novel media for "clean" liquid-liquid extraction. *Chem. Commun.* **1998**, 1765–1766. [CrossRef]
6. Zhao, H.; Malhotra, S.V. Applications of ionic liquids in organic synthesis. *Aldrichimica Acta* **2002**, *35*, 75–83. [CrossRef]
7. De los Rios, A.P.; Hernandez-Fernandez, F.J.; Lozano, L.J.; Sanchez, S.; Moreno, J.I.; Godinez, C. Removal of metal ions from aqueous solutions by extraction with ionic liquids. *J. Chem. Eng. Data* **2010**, *55*, 605–608. [CrossRef]
8. Papaiconomou, N.; Lee, J.M.; Salminen, J.; Von Stosch, M.; Prausnitz, J.M. Selective extraction of copper, mercury, silver, and palladium ions from water using hydrophobic ionic liquids. *Ind. Eng. Chem. Res.* **2008**, *47*, 5080–5086. [CrossRef]
9. Zhao, H.; Xia, S.; Ma, P. Use of ionic liquids as 'green' solvents for extractions. *J. Chem. Technol. Biotechnol.* **2005**, *80*, 1089–1096. [CrossRef]
10. Reyna Gonzales, J.M.; Torriero, A.A.; Siriwardana, A.I.; Burgar, I.M.; Bond, A.M. Extraction of copper(II) ions from aqueous solutions with a methimazole-based ionic liquid. *Anal. Chem.* **2010**, *82*, 7691–7698. [CrossRef] [PubMed]
11. Germani, R.; Mancini, M.V.; Savelli, G.; Spreti, N. Mercury extraction by ionic liquids: Temperature and alkyl chain length effect. *Tetrahedron Lett.* **2007**, *48*, 1767–1769. [CrossRef]
12. Kidani, K.; Hirayama, N.; Imura, H. Extraction behavior of divalent metal cations in ionic liquid chelate extraction systems using 1-alkyl-3-methylimidazolium bis(trifluoromethanesulfonyl)imides and thenoyltrifluoroacetone. *Anal. Sci.* **2008**, *24*, 1251–1254. [CrossRef] [PubMed]
13. Egorov, V.; Djigailo, D.I.; Momotenko, D.S.; Chernyshow, D.V.; Torochesnikova, I.I.; Smirnova, S.V.; Pletner, I.V. Task-specific ionic liquid trioctylmethylammonium salicylate as extraction solvent fortransition metal ions. *Talanta* **2010**, *80*, 1177–1182. [CrossRef] [PubMed]
14. Lertlapwasin, R.N.; Bhawawet, N.A.; Imyin, A.S.; Fuangswasdi, S. Ionic liquid extraction of heavy metal ions by 2-aminothiophenol in 1-butyl-3-methylimidazolium hexafluorophosphate and their association constants. *Sep. Purif. Technol.* **2010**, *72*, 70–76. [CrossRef]

15. Holbrey, J.D.; Visser, A.E.; Spear, S.K.; Reichert, W.M.; Swatloski, R.P.; Roger, R.D. Mercury(II) partitioning from aqueous solutions with a new, hydrophilic ethylene-glycol functionalized *bis*(methylimidazolium) ionic liquid. *Green Chem.* **2003**, *5*, 129–135. [CrossRef]
16. Visser, A.E.; Swatloski, R.P.; Reichert, W.M.; Mayton, R.; Sheff, S.; Wierzbicki, A.; Davis, J.H., Jr; Rogers, R.D. Task-specific ionic liquids for the extraction of metal ions from aqueous solutions. *Chem. Commun.* **2001**, 135–136. [CrossRef]
17. Visser, A.E.; Swatloski, R.P.; Reichert, W.M.; Mayton, R.; Sheff, S.; Wierzbicki, A.; Davis, J.H., Jr.; Rogers, R.D. Task-specific ionic liquids incorporating novel cations for the coordination and extraction of Hg^{2+} and Cd^{2+}: Synthesis, characterization, and extraction studies. *Environ. Sci. Technol.* **2002**, *36*, 2523–2529. [CrossRef] [PubMed]
18. Harjani, J.R.; Friscic, T.; MacGillivray, L.R.; Singer, R.D. Removal of metal ions from aqueous solutions using chelating task specific ionic liquids. *Dalton Trans.* **2008**, *34*, 4595–4601. [CrossRef]
19. Olivier, J.H.; Camerel, F.; Selb, J.; Retailleau, P.; Ziessel, R. Terpyridine-functionalized imidazolium ionic liquids. *Chem. Commun.* **2009**, 1133–1135. [CrossRef] [PubMed]
20. Jensen, M.P.; Neuefeind, J.; Beitz, J.V.; Skanthakumar, S.; Soderholm, L. Mechanisms of metal ion transfer into room-temperature ionic liquids: The role of anion exchange. *J. Am. Chem. Soc.* **2003**, *125*, 15466–15473. [CrossRef] [PubMed]
21. Mehdi, H.; Binnemans, K.; Van Hecke, K.; Van Meervelt, L.; Nockemann, P. Hydrophobic ionic liquids with strongly coordinating anions. *Chem. Commun.* **2010**, *46*, 234–236. [CrossRef] [PubMed]
22. Jensen, M.P.; Dzielawa, J.A.; Rickert, P.; Dietz, M.L. EXAFS Investigations of the mechanism of facilitated ion transfer into a room temperature ionic liquid. *J. Am. Chem. Soc.* **2002**, *124*, 10664–10665. [CrossRef] [PubMed]
23. Dietz, M.L.; Dzielawa, J.A. Ion-exchange as a mode of cation transfer into room-temperature ionic liquids containing crown ethers: Implications for the 'greenness' of ionic liquids as diluents in liquid–liquid extraction. *Chem. Commun.* **2001**, 2124–2125. [CrossRef]
24. Dietz, M.L.; Dzielawa, J.A.; Laszak, I.; Young, B.A.; Jensen, M.P. Influence of solvent structural variations on the mechanism of facilitated ion transfer into room-temperature ionic liquids. *Green Chem.* **2003**, *6*, 682–685. [CrossRef]
25. Dietz, M.L.; Stepinski, D.C. A ternary mechanism for the facilitated transfer of metal ions into room-temperature ionic liquids (RTILs): Implications for the "greenness" of RTILs as extraction solvent. *Green Chem.* **2005**, *7*, 747–750. [CrossRef]
26. Messadi, A.; Mohamadou, A.; Boudesocque, S.; Dupont, L.; Guillon, E. Task-specific ionic liquid with coordinating anion for heavy metal ion extraction: Cation exchange versus ion-pair extraction. *Sep. Purif. Technol.* **2013**, *107*, 172–178. [CrossRef]
27. Root, A.; Binnemans, K. Efficient separation of transition metals from rare earths by an undiluted phosphonium thiocyanate ionic liquid. *Phys. Chem. Chem. Phys.* **2016**, *18*, 16039–16045. [CrossRef] [PubMed]
28. Parmentier, D.; Van der Hoogerstraete, T.; Banerjee, D.; Valia, Y.A.; Metz S, J.; Binnemans, K.; Kroon, M.C. A mechanism for solvent extraction of first row transition metals from chloride media with the ionic liquid tetraoctylammonium oleate. *Dalton Trans.* **2016**, *45*, 9661–9668. [CrossRef] [PubMed]
29. Khalil, M.; Radalla, A. Binary and ternary complexes of inosine. *Talanta* **1998**, *46*, 53–61. [CrossRef]
30. Baran, E.J. The saccharinate anion: A versatile and fascinating ligand in coordination chemistry. *Quim. Nova* **2005**, *28*, 326–328. [CrossRef]
31. Martın, S.; Gotzone Barandika, S.M.; Ruiz de Larramendi, J.L.; Corte, R.; Font-Bardia, M.; Lezama, L.; Serrna, Z.E.; Solans, X. Structural analysis and magnetic properties of the 1D and 3D compounds, [Mn(dca)2$n$bipym] (M = Mn, Cu; dca = Dicyanamide; bipym = Bipyrimidine; n = 1, 2). *Inorg. Chem.* **2001**, *40*, 3687–3692.
32. Mohamadou, A.; Van Albada, G.A.; Kooijman, H.; Wieczorek, B.; Spek, A.L.; Reedijk, J. The binding mode of the ambidentate ligand dicyanamide to transition metal ions can be tuned by bisimidazoline ligands with H-bonding donor property at the rear side of the ligand. *N. J. Chem.* **2003**, *27*, 983–988. [CrossRef]
33. Boudesocque, S.; Mohamadou, A.; Martinez, A.; Déchamps, I.; Dupont, L. Use of dicyanamide ionic liquids for metal ions extraction. *RSC Adv.* **2016**, *6*, 107894–107904. [CrossRef]

34. Zhou, Y.; Boudesocque, S.; Mohamadou, A.; Dupont, L. Extraction of metal ions with Task-specific ionic liquids : Influence of a coordinating anion extraction. *Sep. Sci. Technol.* **2015**, *50*, 38–44. [CrossRef]
35. De Gaetano, Y.; Mohamadou, A.; Boudesocque, S.; Hubert, J.; Plantier-Royon, R.; Dupont, L. Ionic liquids derived from esters of Glycine Betaine: Synthesis and characterization. *J. Mol. Liq.* **2016**, *207*, 60–66. [CrossRef]
36. Powell, K.J. SC Query SC-database, data version 4.44, soft version 5.3 46.
37. Mimoun, M.; Pointud, Y.; Juillard, J.; Juillard, J. Interactions in methanol of divalent metal cations with bacterial iono-phores, lasalocid and monensin-Thermodynamical aspects. *Bull. Soc. Chim. Fr.* **1994**, *131*, 13158–13165.
38. Lever, A.B.P. *Inorganic Electronic Spectroscopy*, 2nd ed.; Elsevier: New York, NY, USA, 1986.
39. Skopenko, V.V.; Samoilenko, V.M.; Movchan, O.G. Study of cadmium halide and pseudohalide complexes in dimethylacetamide. *Zh. Neorg. Khim.* **1981**, *26*, 1319–1323.
40. Skopenko, V.V.; Samoilenko, V.M.; Garbous, S. Potentiometric study of lead halide and pseudohalide complexes in dimethylacetamide. *Zh. Neorg. Khim.* **1982**, *27*, 665–668.
41. Valle-Bourrouet, G.; Pineda L, W.; Falvello L, R.; Lusar, R.; Weyhermueller, T. Synthesis, structure and spectroscopic characterization of Ni(II), Co(II), Cu(II) and Zn(II) complexes with saccharinate and pyrazole. *Polyhedron* **2007**, *26*, 4440–4478. [CrossRef]
42. Ali, M.A.; Mirza, A.H.; Ting, W.Y.; Hamid, M.H.S.A.; Bernhardt, P.V.; Butcher, R.J. Mixed-ligand nickel(II) and copper(II) complexes of tridentate ONS and NNS ligands derived from S-alkyldithiocarbazates with the saccharinate ion as a co-ligand. *Polyhedron* **2012**, *48*, 167–173. [CrossRef]

© 2018 by the authors. Licensee MDPI, Basel, Switzerland. This article is an open access article distributed under the terms and conditions of the Creative Commons Attribution (CC BY) license (http://creativecommons.org/licenses/by/4.0/).

Review

Refining Approaches in the Platinum Group Metal Processing Value Chain—A Review

Pia Sinisalo and Mari Lundström *

Aalto University, School of Chemical Engineering, 02150 Espoo, Finland; pia.sinisalo@aalto.fi
* Correspondence: mari.lundstrom@aalto.fi; Tel.: +358-40-487-3434

Received: 28 February 2018; Accepted: 20 March 2018; Published: 22 March 2018

Abstract: Mineable platinum group metal (PGM) deposits are rare and found in relatively few areas of the world. At the same time, the use of PGM is predicted to expand in green technology and energy applications, and PGMs are consequently currently listed as European Union critical metals. Increased mineralogical complexity, lower grade ores, and recent PGM production expansions give rise to the evaluation of the value chain of the capital-intensive conventional matte smelting treatment and other processing possibilities of the ore. This article will review the processes and value chain developed to treat ores for PGM recovery, highlighting hydrometallurgical refining approaches. It groups processes according to their rationale and discusses the special features of each group.

Keywords: platinum group metals; value chain; refining; leaching

1. Introduction

The platinum group metals (PGMs) are a family of six metals. They are rare and possess extraordinary physical and chemical properties, e.g., resistance to corrosion and oxidation, electrical conductivity, and catalytic activity. The most economically important of the PGMs are platinum, palladium, and rhodium while ruthenium, iridium, and osmium are less prevalent and less in demand [1].

The demand for PGM is fundamentally strong. The demand for platinum breaks down into four segments: automotive, jewellery, industrial (the chemical and petroleum industries, the dental and medical sectors, the glass industry, and a range of other end uses), and investment [2]. The demand for PGMs has risen significantly after the introduction of catalytic converters to control vehicle exhaust emissions, with the catalyst application accounting for a 6% market share for platinum and 16% for ruthenium in 2012 [3].

Over the last five years, between 72% and 78% of total annual platinum supply has come from primary mining output [4] and the rest originates from secondary materials such as spent catalysts and electronic scrap. Most of the current primary PGM production is derived from sulphide ores that typically also contain nickel, copper, and other metals. PGM annual production amounts to around 400 t, [5] of which platinum accounts for approximately 192 t and ruthenium approximately 12 t [6].

Grandell et al. [6] have predicted a path for the global energy system with an assumed annual 2% increase in global gross domestic product, resulting in 290% and 73% cumulative consumption of the world's known reserves of platinum and ruthenium, respectively by 2050. This highlights the critical nature of this metal group and justifies the increasing focus on the primary and secondary processing of PGMs. This article will discuss primary PGM production.

2. Regional and Mineralogical Aspects of PGM Production

According to Jones [1], PGMs are commonly associated with nickel-copper sulphides in magmatic rocks. PGMs are produced either as primary products or by-products of the nickel and copper, depending on the relative concentrations of the metals in the ore. Cole and Ferron [7] also describe ores in which PGM values are noted but not recovered due to low concentrations or recovered as a by-product with little or no economic advantage for the primary producer. PGMs originating from these sources are for instance recovered from copper refineries as a by-product in the United States and Canada.

Mineable PGM deposits are very rare and found in relatively few areas of the world. South Africa dominates PGM world production with 58%. Russia accounts for a further 26%, most of this as a co-product of nickel mining. Nearly all of the rest comes from Zimbabwe, Canada, and the United States. One third of PGMs are produced as co-products of nickel mining [5].

The largest known PGM deposit, the Bushveld Complex in South Africa, contains more than two thirds of the world's reserves of PGMs. The Great Dyke in Zimbabwe is the second largest known deposit of platinum. Other primary PGM-rich deposits include the Stillwater deposit of the United States and the Lac des Isles deposit of Canada. PGMs are produced in significant quantities as by-products from the Norilsk-Talnakh area of Russia and the Sudbury deposit of Canada [1]. Other deposits occur e.g., in Finland and China. Most deposits are fairly small, less than 100 million tons of ore [8].

PGMs have diverse associations and the primary minerals associated with PGMs are pyrrhotite, chalcopyrite, and pentlandite. Chalcopyrite and pentlandite are generally well recovered, as are any PGMs in their lattice or present in any platinum group mineral blebs that they may contain. PGMs associated with pyrrhotite are more challenging to recover and may be largely rejected at the mine site, e.g., in the Sudbury basin, to minimize operating costs and environmental pollution [9].

Liddell and Adams [10] state that pentlandite, chalcopyrite, and cobaltiferrous pyrite are commonly associated with PG mineralisation and that pyrrhotite is also seen in PGM ores. They point out that PG minerals present in the ore have varying associations with base metal sulphide minerals and gangue minerals. A move to increasingly finer grind sizes has occurred as a response to the trend of maximising recovery. Fine gangue particles tend to report to the concentrate more readily than coarse gangue particles.

Gold is often associated with PGM deposits and is treated as part of a family together with platinum, palladium, and rhodium, collectively known as 4E. Nickel and copper are the most prevalent base metals in sulphide ores containing PGMs. Chromite (used to derive chrome) is another significant by-product while cobalt, silver, selenium, and tellurium are found and recovered in trace quantities [11]. Most deposits contain 1–5 g/t 4E [8].

The International Mineralogical Association has recognized nearly 110 different types of platinum group carrying minerals. PGMs form a number of minerals ranging from sulphides to tellurides, antimonides to arsenides, and alloys to native metals [9].

3. PGM Processing and Its Value Chain

PGM processing consists principally of four steps (Figure 1). PGMs are recovered through underground and open pit mining from poly-metallic sulphide ores [11]. They are initially concentrated by flotation. Conventionally, flotation concentrates are enriched by smelting, i.e., iron is combined with silica and lime, oxidised, and separated as a slag leaving the sulphide minerals in a separate matte phase. The smelter can be a base metal smelter or a PGM smelter [12]. The matte is refined using reverse leaching in which most of the accompanying minerals such as copper, nickel, cobalt, and any residual iron sulphides are dissolved. The leach solutions are processed to recover the base metals. In the case of the PGM smelter and the base metal smelter, an upgraded PGM concentrate is left behind or PGMs eventually find their way into the anodic slimes, respectively. PGM concentrate and anodic slimes are further treated at the PGM refinery or precious metal refinery/plant to recover the PGMs and precious metals [7].

In the case of the PGM smelter, electric furnaces are typically used and operated at 1350 °C. Temperatures of up to 1600 °C may be necessary for chromite-containing concentrates owing to the higher content of chromium and magnesium oxides. The furnace matte is further processed by converting, i.e., air is blown into the molten charge to oxidise and remove the iron and its associated sulphur. The converter matte is typically granulated. Anglo Platinum slow-cools the matte, which results in PGM concentrate on the grain boundaries and tends to form ferromagnetic species that are recovered by magnetic separation. The ferromagnetic fraction is treated to dissolve the associated base metals leaving an enriched PGM concentrate for the precious metal refinery [7].

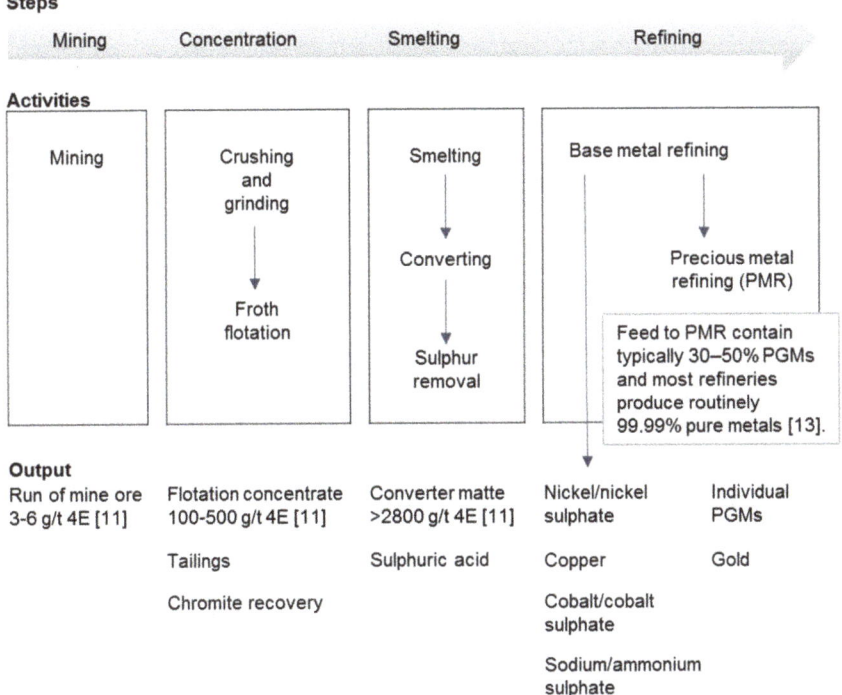

Figure 1. Simplified platinum group metal (PGM) processing chain (conventional matte smelting treatment). 4E = Pt, Pd, Rh, and Au. Adapted from Ndlovu [11] and Cramer [13].

The four major PGM mining companies, namely Anglo Platinum, Norilsk, Impala Platinum, and Lonmin Platinum, all South African apart from the Russian Norilsk, are integrated from mining to refining and account for 80% of the market [14]. One of them, Lonmin, outsources the refining of excess base metals [11]. According to Cramer [15], their nickel production is sold to other nickel refiners as nickel sulphate. Levine et al. [16] state that Norilsk outsources PGM-containing copper solids at the Harjavalta refinery. Norilsk relies on third party PGM refiners according to Ndlovu [11].

Cramer [13] stated in 2001 that several of the South African PGM producers reviewed their refining capacities in order to align their base metals capacities with the major PGM production expansions planned over the next several years. Anglo Platinum found that PGM processing capacity (for PGM dominant ores) was not easily globally traded, with very few concentrated alternatives, namely other integrated PGM processing facilities. Their choice was to keep deploying capacity downstream [11]. In 2011, an expansion project at Anglo Platinum's base metals refinery in Rustenburg was completed and operational [17].

Cramer [15] stated in 2008 that a couple of dozen minor platinum mining companies were being set up in Southern Africa. Most of them intended to mine, concentrate the PG minerals to flotation concentrate, and sell the product. There was also several ongoing studies, in particular smelter studies, on further upgrading the product as another intermediate and saleable product. Further investment in process capacity was often considered in conjunction with other smaller mining companies faced with limited capacity for concentrate smelting and base metal refining within the local platinum industry or together with one of the major platinum companies.

The major factors of process value come down to capital and operating efficiency. Generally, the focus to drive value is largely in the mining and concentration steps in which the most losses are incurred. For example, total losses in the mining step before the process, account for 27% compared with 15% overall processing losses, of which the concentration step accounts for 13% at an underground platinum mine. Non-integrated mining companies produce approximately 10% of the major PGM mining companies' processing input. Of the operating costs, smelting and refining account for 15% (Table 1). [11] The capital investments to process PGMs from flotation concentrate are very large and a significant barrier to entry for smaller mining companies [15].

Table 1. Value chain for PGM processing (adapted from Ndlovu [11]).

Economics	Mining	Concentration	Smelting	BMR	PMR
Purchase price	50% of ref value	85%	90%	90%	95–98%
Opex distribution	67%	20%	8%	4%	2%

BMR = base metal refining; PMR = precious metal refining.

Platinum producers evaluate process routes and their value chain alternative to conventional matte smelting treatment, which is presented in Figure 1. Projects are evaluated for cost-effective processing with low-grade ores and in some cases higher content of deleterious elements such as chromium [18]. Hydrometallurgical plants are best suited for small operations and this is the direction usually taken [7]. Smelters have a high capital cost and favour large capacities. In addition, sulphur dioxide generation at the smelters requires conversion to sulphuric acid due to increasingly stringent regulations, which have a large negative value in the overall economics [19] and a considerably positive environmental impact.

4. Base Metal Refining

According to Cole and Ferron [7], all PGM base metal refineries in the western world use Sherrit-Gordon's sulphuric acid pressure leach process. The process flow sheets vary in each operation. The now closed Hartley Platinum operations applied the Outokumpu process. Both processes have the same rationale and the chemistry underlying both processes is the same. In general, the same processes used in nickel and copper-nickel refineries are also used for PGM base metal refineries. The following focuses on the latter.

Finely ground matte (or the non-magnetic matte at the Anglo Platinum operation) is leached in a number of stages under increasingly more oxidizing conditions to produce a residue containing PGMs and typically nickel- and copper-rich solutions. Copper is recovered as electrowon cathode and nickel as metal powder by hydrogen reduction or electrowon cathode. At the Stillwater refinery, a bulk nickel-copper solution is produced and shipped to an outside nickel refinery. The residue can be subjected to further treatment to upgrade the PGM content before the final residue passes to the PGM refinery [7].

A typical example is the process used at Rustenburg Base Metals Refiners that treats the non-magnetic nickel-copper matte from a slow cooling matte separation process (Figure 2). The initial stage, i.e., copper removal, operates under mild conditions to leach some nickel and cobalt and aims to precipitate copper from the nickel sulphate solution with fresh matte. The primary leach stage aims to dissolve all the nickel and some copper from the matte while the secondary leach stage aims to dissolve all the remaining copper and some iron from the matte [7].

Fresh matte consumes the available acid and iron precipitates during the decopperisation. Iron can also be removed in a special high temperature step prior to copper removal. The neutralised iron-

and copper-free nickel solution is suitable for the recovery of nickel and cobalt products. The copper solution undergoes a selenium removal stage that also removes any co-dissolved PGM prior to electrowinning [7].

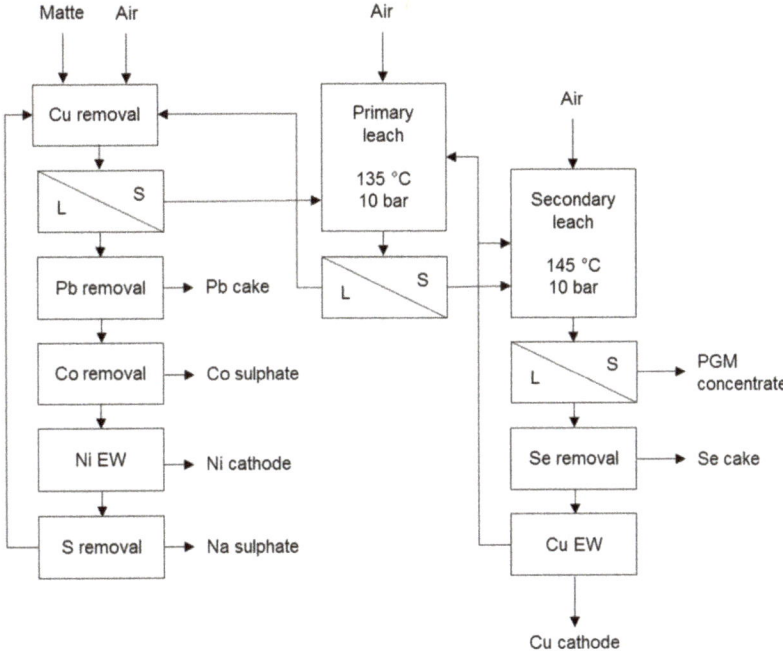

Figure 2. Simplified flow sheet of Rustenburg Base Metals Refinery (adapted from Hofirek & Halton [20]).

Liddell et al. [21] also present the Falconbridge process for the removal of base metals from matte (Figure 3). The nickel is dissolved with hydrochloric acid in a non-oxidative leach and the solution is purified, i.e., sulphur is removed by oxidation, and iron and cobalt by solvent extraction. The nickel is then recovered and the acid regenerated. The leach residue is roasted and copper is then leached with sulphuric acid. The copper is electrowon from the solution and the residue passes to the PGM refinery. PGMs are a by-product of mostly nickel, copper, and cobalt in this process according to Cole & Ferron [7].

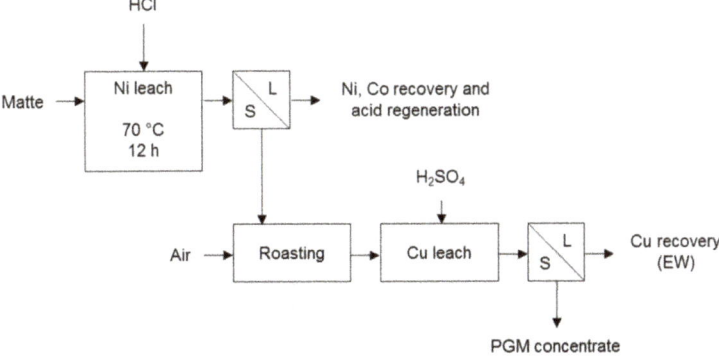

Figure 3. Falconbridge method for base metal removal (adapted from Thornhill et al. [22]).

5. PGM Refining

The individual PGMs are refined totally or partially using a classical refining process with solvent extraction and molecular recognition developments. The classical refining process is based on PGM chloride chemistry. It involves successive steps of precipitation and redissolution to purify the metals, followed by thermal reduction to metal. However, it gives a poor first time yield for refined metals and requires lengthy refining times, e.g., up to six months for rhodium [7].

Crundwell et al. [23] classify PGM refining processes by the technique used to separate platinum and palladium from one another. Of the major South African PGM mining companies, Lonmin Platinum uses the precipitation process, Anglo Platinum the solvent extraction process, and Impala Platinum the ion-exchange process (Table 2). The solvent extraction process is also used, by Johnson Matthey and Vale for instance.

Table 2. Extraction methods in a few PGM refineries (adapted from Crundwell et al. [23]).

Process Details	Lonmin Platinum	Anglo Platinum	Impala Platinum
Technology	Precipitation	Solvent extraction, SX	Ion exchange, IX
Feed	65–75% PGMs	30–50% PGMs	60–65% PGMs
Dissolution	HCl/Cl_2 at 65 °C, atm	HCl/Cl_2 at 120 °C, 4 bar	HCl/Cl_2 at 85 °C, 1 bar
Extraction order *	Au, Ru, Pt, Pd	Au, Pd, Pt, Ru	Au, Pd, Ru, Pt
Gold extraction	Crude sponge produced by hydrazine reduction.	SX with MIBK, reduction with oxalic acid.	IX, reduction with hydroquinone.
Palladium extraction	Precipitation with ammonium acetate.	SX with β-hydroxyl oxime, stripping with ammonium hydroxide.	IX with SuperLig 2, stripping with ammonium bisulphate and precipitation with ammonium hydroxide.
Platinum extraction	Precipitation with ammonium chloride.	SX with tri-n-octylamine, stripped with HCl, and precipitated with ammonium chloride.	Precipitation with ammonium chloride.
Ruthenium extraction	Distillation with sodium chlorate and bromate.	Distillation with sodium chlorate and bromate.	Distillation with sodium chlorate and bromate.
Iridium extraction	Precipitation with ammonium chloride.	SX with n-iso-tridecyl tri-decanamide.	IX, precipitation with ammonium chloride.
Rhodium extraction	Precipitation as $Rh(NH_3)_5Cl_3$, dissolution, precipitation with ammonium chloride.	IX followed by precipitation with diethylene tri-ammine.	Precipitation with an organic ammine.

MIBK = methyl isobutyl ketone. * Rh and Ir separated from solution with hydrolysis for refining after Ru distillation.

6. Non-Smelter-Based Processes

Non-smelter-based processes developed for PGM extraction typically include sequential base metal leach and PGM leach steps or leaching the PGMs together with base metals. Both approaches are followed by solution purification operations for the recovery of metal or intermediate products.

The processes include sulphide treatment to break down the sulphide matrix to liberate the base and PG metals locked in the matrix. Consequently, process flow sheets commonly comprise at least roasting, pressure oxidation, or fine grinding prior to leaching, or the matrix is attacked by bio or chemical oxidation during leaching. Pressure oxidation is used at conventional or mild temperatures to completely oxidise sulphides to sulphate or to oxidise them predominantly to sulphur, respectively. The latter requires a finer feed according to Milbourne et al. [12]. This may be done by ultrafine grinding, which produces particles sized within the 1 to 20 um range. Small enough particles allow disintegration of the leached mineral before the sulphur layer becomes thick enough to passivate it [8].

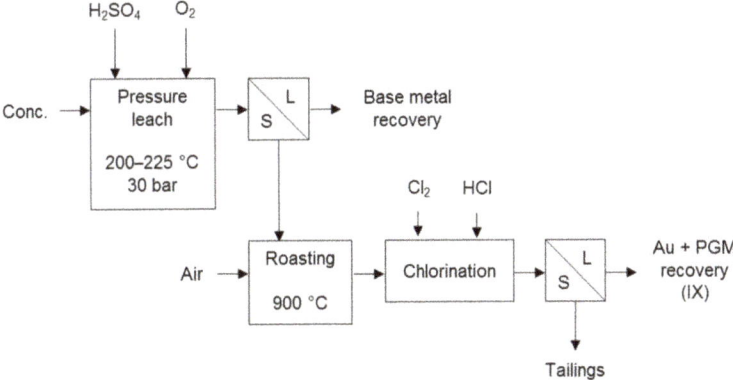

Figure 4. Block diagram of the Kell process (adapted from Liddell & Adams [10]).

The Kell process (Figure 4) first selectively removes base metals and sulphur by pressure oxidation. The residue is subjected to roasting to render the PGMs amenable to recovery by chlorination. The conventional pressure oxidation of the concentrate converts all the sulphur to sulphate, avoiding sulphur dioxide in the roaster off-gas. This process was developed to treat PGM-bearing concentrates and has been successively tested on several concentrates [10], including Platreef concentrate from the Bushveld Complex of South Africa [18].

Processes that co-dissolve PG and base metals in a sulphate environment make use of halogens to promote PGM dissolution. These processes include the TML process (Figure 5) and the Platsol process (Figure 6). The former was developed to recover PGM from the oxidised portion of the Hartley deposit in Zimbabwe. The latter was developed to treat the NorthMet deposit in Minnesota, USA and has also been applied to treat other PGM/Cu–Ni concentrates, Cu–Au concentrates, Cu–Ni–PGM matte, Pt laterites and auto-catalysts [7]. Both processes use halogens for the formation of PGM complexes and the TML process also for redox potential control.

Chloride-based processes co-dissolve PG and base metals, promoting dissolution by other means. Cole and Ferron [7] state that the North American Palladium process (Figure 7) was developed for the treatment of Lac des Isles PGM sulphide concentrate. It uses pressure and hydrochloric acid with small amounts of nitric acid. Soluble nitrogen species transport oxygen to the surface of the solid particle, which enhances the rate of dissolution of metal sulphides. The nitric acid is continuously regenerated by the oxygen gas applied. Partial oxidising roasting was included in the flow sheet, as it increased the platinum recovery significantly [24].

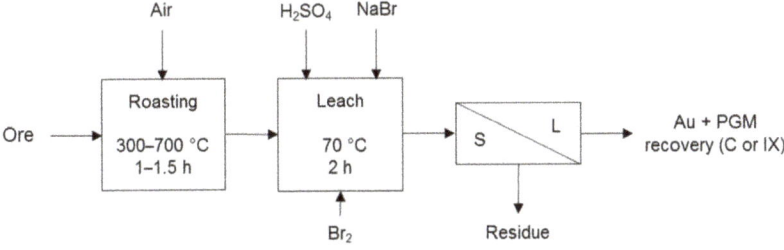

Figure 5. Simplified TML process diagram for Hartley oxide ore. Other leach parameters were 100 g/L H_2SO_4, 10 g/L NaBr, >800 mV ORP with Br_2 using Geobrom 3400 as oxidant, and 24–40% solids. Adapted from Cole and Ferron [7]. Geobrom 3400 is a sodium bromide solution in which liquid bromine is dissolved.

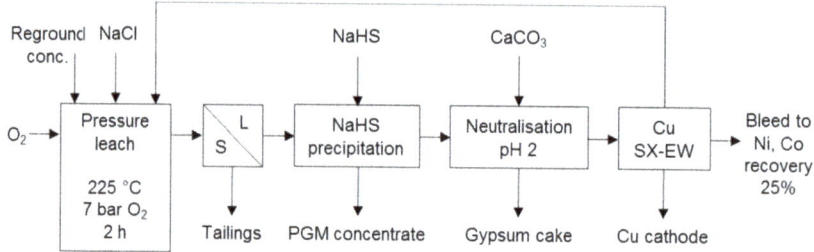

Figure 6. Flow sheet for the Platsol treatment of NorthMet concentrate. NaCl addition was 5–20 g/L. Adapted from Cole and Ferron [7].

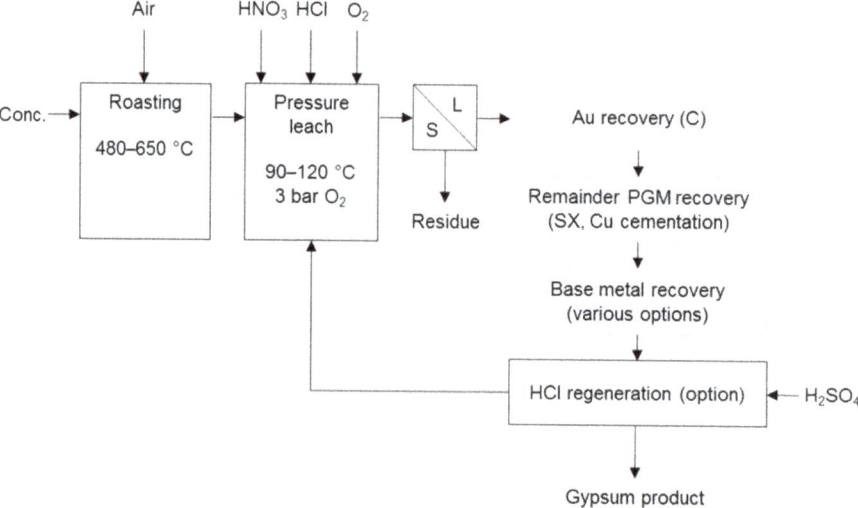

Figure 7. Block diagram for the treatment of the Lac des Iles PGM concentrate (adapted from McDoulett & Reschke [24]).

Several chloride-based or chloride-aided sulphate processes such as the CESL [25], Outotec [26] and Intec [19] copper processes and HydroCopper [27] have been developed to treat copper sulphide concentrates. They exclude PGMs from consideration. This is also the case for the chloride-based Outotec gold process [28] developed to treat gold-bearing raw materials. Based on the literature, they have not been applied to treat PGM-dominant concentrates apart from some laboratory batch tests carried out with the Intec process and mentioned by Milbourne et al. [12].

Other direct leaching processes for the treatment of chalcopyrite copper concentrate have also been developed. These include the nitrogen species catalysed process and several other sulphate-based processes. Typically none focus on PGM recovery. Milbourne et al. [12] reviewed and theoretically discussed the dissolution and deportment of PGMs in many of these processes. Mpinga et al. [8] also reviewed these processes. However, test work is required to determine the behaviour of PGMs and the method of recovery among other process modifications needed. We will concentrate on reviewing the tested technology for PGM extraction.

Cyanide-based processes use elevated temperature or pressure during cyanidation to promote PGM dissolution. The processes that co-dissolve PG and base metals were apparently mainly developed to process oxide material with low copper concentrations, as the presence of sulphide minerals and high concentrations of copper in the cyanidation feed typically have an adverse effect on cyanidation.

Processes that use cyanidation following base metal leach in a sulphate environment include the two-stage selective pressure leach process (Figure 8) and sequential heap leach process (Figure 9). The former has been tested for treating Jinbaoshan mine's concentrate from China [29] and the latter for treating Platreef ore and its low-grade concentrate [30]. Conventional pressure oxidation avoids and bioleaching decreases the formation of ineffective thiocyanates in these processes.

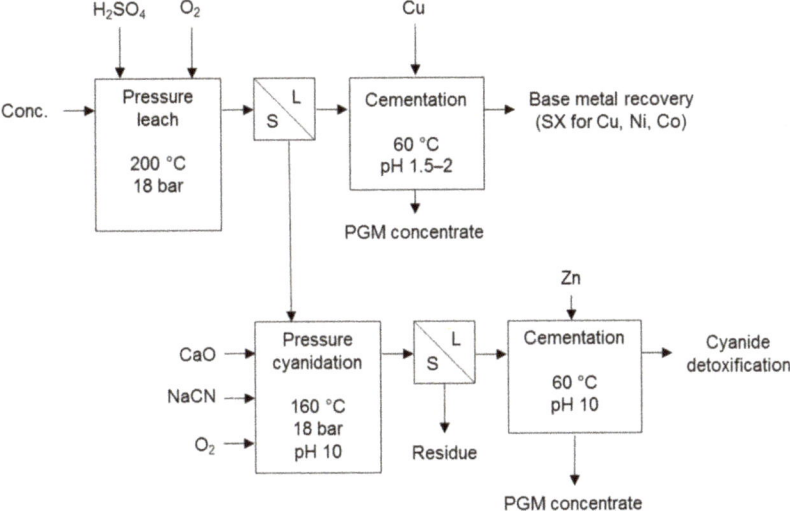

Figure 8. Two-stage selective pressure leaching process diagram (adapted from Huang et al. [29]).

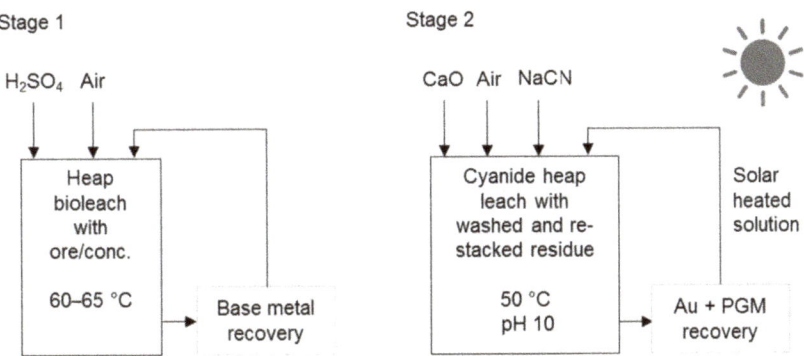

Figure 9. Schematic diagram of the sequential heap leach process (adapted from Mwase et al. [30]).

Cyanidation processes co-dissolving PG and base metals were developed to treat Coronation Hill ore (Figure 10) [7] and Panton flotation concentrate (Figure 11) in Australia. The Panton process comprises cyanidation following calcination and the dissolved metals are co-precipitated. The precipitate is the final product or is upgraded to PGM concentrate and a separate base metal concentrate [31]. The current focus of the project is away from the Panton process [32].

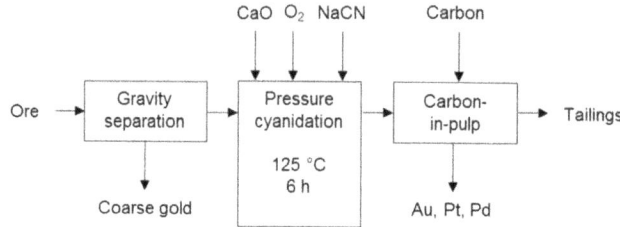

Figure 10. Cyanidation process diagram for Coronation Hill ore (adapted from Cole & Ferron [7]).

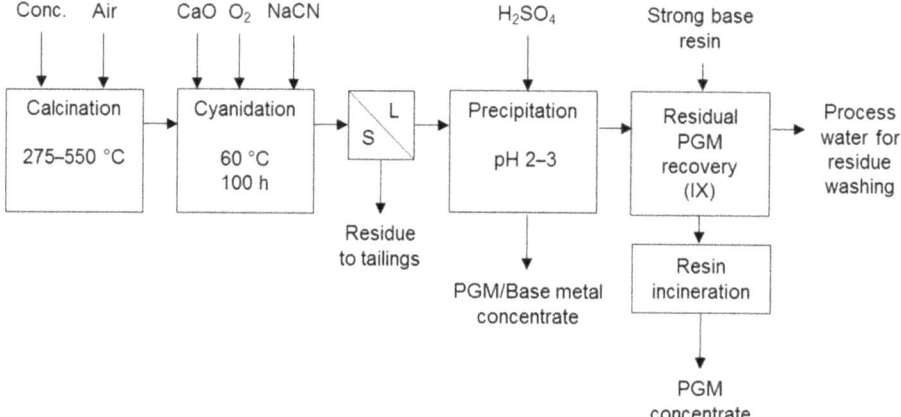

Figure 11. Panton process flow sheet (adapted from Lewins & Greenaway [31]).

7. Conclusions

Due to the critical nature of PGMs and the increasing need for clean technology applications, the recovery of PGMs from both primary and secondary raw materials is of increasing importance. The current article focused on the PGM processes of the former and their value chain, highlighting hydrometallurgical refining approaches.

PGMs are recovered from low-grade resources and from a variety of host minerals due to their high value. Annual production of PGMs is around 400 t, with Anglo Platinum having the biggest share. One third of PGMs are produced as co-products of nickel mining. More than 80% of PGMs originate from two countries, namely South Africa and Russia. The former produces PGMs as primary products and the latter mainly as by-products of nickel. The geographical concentration of the resources increases the supply risk of this metal group.

The major four PGM mining companies are integrated from mining to refining. They account for 80% of the market; conventional matte smelting treatment being the prevailing processing method. Capital investments to process PGMs from flotation concentrate are very large and a significant barrier to entry for smaller mining companies. They typically produce flotation concentrate and sell the product for further refining. Other solutions for their value chain may emerge as a result of several studies—these may lead to joint ventures in the near future.

Alternative process routes are typically evaluated when the conventional method is not cost-effective. The direction is usually towards hydrometallurgical plants, as they are best suited for small operations. The technology developed makes use of metallurgy diversely and takes into account a specific raw material. Consequently, the proposed processes differ from each other. The proposed unit operations are predominantly proven and in common use. The chemistry underlying them is generally known in quite a detail. The suggested leaching media are sulphide-, chloride- or cyanide-based. These processes have a lack of industrial reference, which make it more challenging for them to gain a foothold.

Acknowledgments: The authors greatly acknowledge the Circular Metal Ecosystem (CMEco) project (7405-31-2016) for funding the research.

Author Contributions: Pia Sinisalo planned the content of the paper, conducted the literature research and the majority of the manuscript writing; Mari Lundström participated in the writing and was the instructor and supervisor of the work.

Conflicts of Interest: The authors declare no conflict of interest.

References

1. Jones, R.T. An overview of Southern African PGM smelting. In *Nickel and Cobalt 2005, Challenges in Extraction and Production*; Donald, J., Shoneville, R., Eds.; Canadian Institute of Mining, Metallurgy and Petroleum: Montreal, Quebec, Canada, 2005; pp. 147–178.
2. Jollie, D. Forecasting Platinum Supply and Demand. 2016. Available online: https://www.platinuminvestment.com/files/Platinum_Fundamentals_forecast_Glaux_2016.pdf (accessed on 31 May 2017).
3. Butler, J. Platinum 2012. Available online: http://www.platinum.matthey.com/publications/pgm-market-reviews/market-review-archive/platinum-2012 (accessed on 16 November 2017).
4. Platinum Mining Supply Predicted to Fall Further in 2017. World Platinum Investment Council Press Release, 15 May 2017, 1–3. Available online: https://www.platinuminvestment.com/files/224537/WPIC_PR_PQ%20Q1%202017_15052017.pdf (accessed on 9 February 2018).
5. The Primary Production of Platinum Group Metals (PGMs). International Platinum Group Metals Association Fact Sheet, 1–7. Available online: http://ipa-news.com/assets/sustainability/Primary%20Production%20Fact%20Sheet_LR.pdf (accessed on 9 February 2018).
6. Grandell, L.; Lehtilä, A.; Kivinen, M.; Koljonen, T.; Kihlman, S.; Lauri, L.S. Role of critical metals in the future markets of clean energy technologies. *Renew. Energy* **2016**, *95*, 53–62. [CrossRef]
7. Cole, S.; Ferron, J. A Review of the Beneficiation and Extractive Metallurgy of the Platinum Group Elements, Highlighting Recent Process Innovations. SGS Minerals Services Technical Paper 2002-03, 1–43. Available online: http://www.sgs.com/-/media/global/documents/technical-documents/sgs-technical-papers/sgs-min-tp2002-03-beneficiation-and-extractive-metallurgy-of-pge.pdf (accessed on 9 February 2018).
8. Mpinga, C.N.; Eksteen, J.J.; Aldrich, C.; Dyer, L. Direct leach approaches to platinum group metal (PGM) ores and concentrates: A review. *Miner. Eng.* **2015**, *78*, 93–113. [CrossRef]
9. Xiao, Z.; Laplante, A.R. Characterizing and recovering the platinum group minerals—A review. *Miner. Eng.* **2004**, *17*, 961–979. [CrossRef]
10. Liddell, K.S.; Adams, M.D. Kell hydrometallurgical process for extraction of platinum group metals and base metals from flotation concentrates. *J. South. Afr. Inst. Min. Metall.* **2012**, *112*, 31–36.
11. Ndlovu, J. Overview of PGM Processing. Anglo Platinum. Available online: http://www.angloamericanplatinum.com/~/media/Files/A/Anglo-American-Platinum/investor-presentation/standardbankconference-anglo-american-platinum-processing-111114.pdf (accessed on 31 May 2017).
12. Milbourne, J.; Tomlinson, M.; Gormely, L. Use of hydrometallurgy in direct processing of base metal/PGM concentrates. In *Hydrometallurgy 2003*; Young, C.A., Alfantazi, A.M., Anderson, C.G., Dreisinger, D.B., Harris, B., James, A., Eds.; TMS: Pittsburgh, PA, USA, 2003; Volume 1, pp. 617–630.
13. Cramer, L.A. The extractive metallurgy of South Africa's platinum ore. *JOM* **2001**, *53*, 14–18. [CrossRef]
14. Ndlovu, J. Precious Metals Supply. Anglo Platinum, 2015. Available online: https://www.google.fi/url?sa=t&rct=j&q=&esrc=s&source=web&cd=1&cad=rja&uact=8&ved=0ahUKEwjw36qOv7HXAhVREewKHe1OCD8QFgglMAA&url=http%3A%2F%2Fec.europa.eu%2FDocsRoom%2Fdocuments%2F14045%2Fattachments%2F1%2Ftranslations%2Fen%2Frenditions%2Fnative&usg=AOvVaw16kK3iMz6QoKTO3MH90wyl (accessed on 2 June 2017).
15. Cramer, L.A. What is your PGM concentrate worth? In *Third International Platinum Conference 'Platinum in Transformation'*; The Southern African Institute of Mining and Metallurgy: Johannesburg, South Africa, 2008; pp. 387–394.

16. Levine, R.M.; Brininstool, M.; Wallace, G.J. The mineral industry of Russia—Strategic information. In *Russia and Newly Independent States (NIS), Mineral Industry Handbook*; International Business Publications: Washington DC, USA, 2015; Volume 1, pp. 40–61.
17. Goosen, S. Rustenburg Base Metals Refinery Completed on Time and on Budget. Mining Weekly, 16 December 2011. Available online: http://www.miningweekly.com/article/rustenburg-base-metals-refinery-998-complete-2011-12-16 (accessed on 9 February 2018).
18. Adams, M.; Liddell, K.; Holohan, T. Hydrometallurgical processing of Platreef flotation concentrate. *Miner. Eng.* **2011**, *24*, 545–550. [CrossRef]
19. Everett, P.K. Development of Intec Copper Process by an international consortium. In *Hydrometallurgy '94*; Springer: Dordrecht, The Netherlands, 1994; pp. 913–922.
20. Hofirek, Z.; Halton, P. Production of high quality electrowon nickel at Rustenburg Base Metals Refiners (Pty.) Ltd. In *Proceedings of the International Symposium on Electrometallurgical Plant Practice*; Claessens, P.L., Harris, G.B., Eds.; Pergamon Press: New York, NY, USA, 1990; pp. 233–252. [CrossRef]
21. Liddell, K.S.; McRae, L.B.; Dunne, R.C. Process routes for beneficiation of noble metals from Merensky and UG-2 ores. *Mintek Rev.* **1986**, *4*, 33–44.
22. Thornhill, P.G.; Wigstol, E.; Van Weert, G. The Falconbridge matte leach process. *JOM* **1971**, *23*, 13–18. [CrossRef]
23. Crundwell, F.; Moats, M.; Ramachandran, V.; Robinson, T.; Davenport, W. *Extractive Metallurgy of Nickel, Cobalt and Platinum Group Metals*; Elsevier: Oxford, UK, 2011; pp. 489–534. ISBN 978-0-08-096809-4.
24. McDoulett, C.D.; Reschke, G.W. Metal Leaching and Recovery Process. Patent EP 0637635 A2, 8 February 1995.
25. Jones, D.L. Chloride Assisted Hydrometallurgical Extraction of Metal. Patent US 5902474 A, 11 May 1999.
26. Valkama, K.; Sinisalo, P.; Karonen, J.; Hietala, K. Method of Recovering Copper and Precious Metals. Patent WO 2014195586 A1, 11 December 2014.
27. Sinisalo, P.; Tiihonen, M.; Hietala, K. Gold recovery from chalcopyrite concentrates in the HydroCopper® process. In *ALTA 2008 Copper Conference*; ALTA Metallurgical Services: Melbourne, Australia, 2008; pp. 1–8.
28. Miettinen, V.; Ahtiainen, R.; Valkama, K. Method of Preparing a Gold-Containing Solution and Process Arrangement for Recovering Gold and Silver. Patent US 20160068927 A1, 10 March 2016.
29. Huang, K.; Chen, J.; Chen, Y.R.; Zhao, J.C.; Li, Q.W.; Yang, Q.X.; Zhang, Y. Enrichment of platinum group metals (PGMs) by two-stage selective pressure leaching cementation from low-grade Pt-Pd sulfide concentrates. *Metall. Mater. Trans B* **2006**, *37*, 697–701. [CrossRef]
30. Mwase, J.M.; Petersen, J.; Eksteen, J.J. A novel sequential heap leach process for treating crushed Platreef ore. *Hydrometallurgy* **2014**, *141*, 97–104. [CrossRef]
31. Lewins, J.; Greenaway, T. The Panton platinum palladium project. *Aust. Inst. Min. Metall. Bull.* **2004**, 24–34.
32. Ferron, C.J. Recovery of gold as by-product from the base-metals industries. In *Gold Ore Processing: Project Development and Operations*, 2nd ed.; Adams, M.D., Ed.; Elsevier: Amsterdam, The Netherlands, 2016; pp. 831–856.

© 2018 by the authors. Licensee MDPI, Basel, Switzerland. This article is an open access article distributed under the terms and conditions of the Creative Commons Attribution (CC BY) license (http://creativecommons.org/licenses/by/4.0/).

Review

Recovery of Gold from Pregnant Thiosulfate Solutions by the Resin Adsorption Technique

Zhonglin Dong, Tao Jiang, Bin Xu *, Yongbin Yang and Qian Li *

School of Minerals Processing and Bioengineering, Central South University, Changsha 410083, China; dongzhonglincsu@csu.edu.cn (Z.D.); jiangtao@mail.csu.edu.cn (T.J.); ybyangcsu@126.com (Y.Y.)
* Correspondence: xubincsu@csu.edu.cn (B.X.); csuliqian@126.com (Q.L.);
  Tel.: +86-150-849-33770 (B.X.); +86-135-741-99228 (Q.L.)

Received: 14 November 2017; Accepted: 6 December 2017; Published: 12 December 2017

Abstract: This review is devoted to an integrated evaluation of the current use and future development of the resin adsorption technique in gold recovery from pregnant thiosulfate solutions. Comparisons are firstly made with other recovery techniques, including precipitation, activated carbon adsorption, solvent extraction, electrowinning and mesoporous silica adsorption. A detailed discussion about the recent advances of the technique in gold recovery from pregnant thiosulfate solutions is then presented from the aspects of gold adsorption on the resins and gold-loaded resin elution, respectively. On the basis of summarizing the present research, the major limitations of the resin adsorption technique are eventually pointed out and future development will also be prospected.

Keywords: gold recovery; pregnant thiosulfate solutions; resin adsorption technique; competitive adsorption; eluent

---

1. Introduction

Cyanide leaching has been used to leach gold from ores for more than a century because of its simple process, high efficiency and low cost. However, there are increasing public worries over the sustained use of cyanide because it is extremely poisonous and can readily cause serious environmental and health problems. In addition, cyanide leaching shows unsatisfactory performance on refractory gold ores such as those containing copper or "preg-robbing" carbon [1]. So, considerable attention has been paid to non-cyanide techniques. Of the alternative techniques, thiosulfate leaching offers the advantages of non-toxicity, low reagent costs, fast leaching rate and good performance in treating certain refractory gold ores, and thus, has been widely accepted by researchers as the most promising non-cyanide technique to replace conventional cyanidation [2,3].

Nevertheless, the successful commercial application of thiosulfate leaching is still rare up to now, except for the development of an ammonia-free thiosulfate leaching process by Barrick Gold Corporation to treat a carbon-bearing sulfide gold ore pretreated with acidic or alkaline pressure oxidation [4,5] whose simplified process flowsheet is presented in Figure 1. Certain factors can account for this, but one of the primary impediments is the difficulty in recovering gold from pregnant thiosulfate solutions [6]. Gold leaching from its ores using thiosulfate solutions has been extensively investigated and is relatively well understood over the past several decades [7–32]. However, limited studies on gold recovery from its leach solutions have been carried out. In recent years, there have been some attempts to recover gold from pregnant thiosulfate solutions by several recovery techniques whose characteristics are summarized and evaluated in Table 1. These studies have indicated that resin adsorption is more suitable compared with other recovery techniques, including precipitation, activated carbon adsorption, solvent extraction, electrowinning and mesoporous silica adsorption.

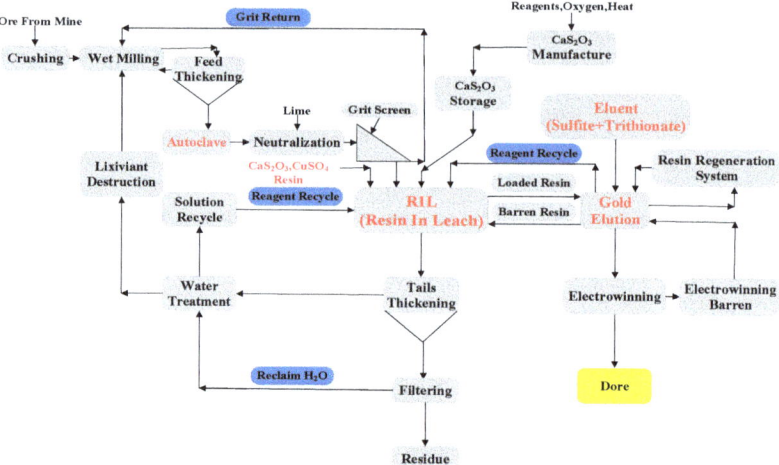

Figure 1. Ammonia-free thiosulfate leaching process by Barrick Gold Corporation to treat a carbon-bearing sulfide gold ore pretreated with acidic or alkaline pressure oxidation [4,5].

Table 1. Evaluation of various gold recovery techniques from pregnant thiosulfate solutions.

Recovery Techinique	Characteristics	References
Precipitation	Technique is simple High consumption of precipitation agents. Low-purity gold product Difficulty in cyclic utilization of pregnant thiosulfate solutions	[33–41]
Activated carbon adsorption	Low requirements on the clarity of solutions Weak affinity for $[Au(S_2O_3)_2]^{3-}$ anion Modification is necessary to improve its gold loading capacity	[42–54]
Solvent extraction	Pregnant thiosulfate solutions with high gold concentration is needed High equipment and operating costs due to complete solid-liquid separation of pulp Dissolution and accumulation of organic extractant is inevitable	[55–63]
Electrowinning	Technique is simple Low current efficiency and high energy consumption due to the undesirable reactions Difficulty in cyclic utilization of pregnant thiosulfate solutions	[64–67]
Mesoporous silica adsorption	High gold loading capacity Separation difficulty between pulp and adsorbent dut to its fine particle size High requirements on the solution pH values	[68–73]
Resin adsorption	Fast adsorption speed and high gold loading capacity Low requirements on the clarity of solutions Simultaneous elution and regeneration at ambient temperature through the elaborate choice of eluent Difficulty in gold elution from gold-loaded resins	[74–80]

Precipitation, also known as the Merrill–Crowe process or cementation, is a common technique for gold recovery from pregnant thiosulfate solutions mainly with metallic powders such as zinc, iron and aluminium [33]. However, additional copper must be added if the solutions are returned to the leaching circuit because copper ions in the solutions are also precipitated out by these metallic powders. Also, these undesirable cations introduce may prevent gold leaching in a thiosulfate leach solution [34]. The use of copper powder is a reasonable choice to avoid the above problems, but the dissolution of copper increases the redox potential of solution and causes the re-dissolution of precipitated gold and enormous oxidation of thiosulfate [35–37]. Furthermore, the dosages of metallic powders needed in precipitation are usually much more than theoretical amounts because the metal surfaces are readily passivated in the solution, therefore resulting in low-purity gold product [38]. In addition, dissolved gold in pregnant thiosulfate solutions can also be reduced by the addition of sodium borohydride and soluble sulfides [39–41]. Unfortunately, the extensive co-precipitation of other metals, particularly copper, in solution also occurs, necessitating further purification of the obtained gold product.

Activated carbon adsorption shows good performance in gold recovery from cyanide solutions because of its high efficiency, moderate cost and high purity of product. However, activated carbon has remarkably less affinity for $[Au(S_2O_3)_2]^{3-}$ anion than $[Au(CN)_2]^-$ anion [42–47], and the potential reasons are the relatively high negative charge of the $[Au(S_2O_3)_2]^{3-}$ anion, steric limitations or specific interactions between the ligand group and carbon active sites [48]. It has been proposed that the affinity order of activated carbon for various gold complexes is: $[Au(SCN)_2]^- > Au[SC(NH_2)_2]_2^+ > [Au(CN)_2]^- \gg [Au(S_2O_3)_2]^{3-}$ [49–51]. The low affinity of activated carbon for $[Au(S_2O_3)_2]^{3-}$ anion makes it difficult to recover gold from pregnant thiosulfate solutions. Two methods can be adopted to improve the recovery technique. One is to add a certain amount of cyanide into the pregnant solutions to produce a more stable gold cyanide complex, followed by adsorption on the activated carbon [52,53]. Another is to modify activated carbon by cyano-cuprous or cupric ferrocyanide [54]. However, the introduction of any cyanide in the two approaches makes the gold recovery process not cyanide-free, which thus cannot be considered as a true alternative to cyanide process.

There have also been some attempts to recover gold from pregnant thiosulfate solutions by solvent extraction technique [55–62]. However, there are several factors limiting its possible commercial application scopes. One is that solvent extraction is suitable for the treatment of clarified solutions containing relatively higher gold concentration than that of resin adsorption. Thus, the complete solid–liquid separation of pulp before extraction is needed, necessitating additional equipment and operating costs. Another problem is that organic extractant can dissolve in the aqueous phase in small amounts causing the losses of extractant [63]. In addition, the accumulation of extractant in the solution is detrimental to its cycle use. Therefore, the technique has not been considered to be economical.

Electrowinning is also an option for gold recovery, and $[Au(S_2O_3)_2]^{3-}$ anions in the pregnant solutions will migrate to the cathode, being reduced to metallic gold [64–67]. However, it is also not a feasible choice because undesirable reactions of other anions including copper (I) thiosulfate complexes, sulfur-oxygen anions and other metal thiosulfate complexes in the solutions can occur on the electrode surface. This not only decreases the current efficiency and increases energy consumption, but also reduces the purity of gold product. Again, the cyclic utilization of pregnant thiosulfate solutions has also become significantly hard due to the irreversible degradation of thiosulfate during electrowinning.

Mesoporous silica, an ordered mesoporous material which presents high adsorption capacity and selectivity for gold complexes, has also recently been suggested as an alternative to recover gold from pregnant thiosulfate solutions [68–72]. The maximum loading capacity reached about 600 mg gold per gram amine-bearing mesoporous silica and an average 80% gold recovery can be obtained. However, the adsorption material is a kind of fine powder whose average particle size is only 1 μm. This limits its practical application to the treatment of pulp due to the separation difficulty between pulp and adsorbent. In addition, it is generally used in neutral condition (pH~7.5) for a longer lifecycle which means that it may be not suitable for gold recovery from pregnant thiosulfate solutions whose common pH range is 9–11 [73].

Compared with the above-mentioned techniques, resin adsorption stands out as the most promising gold recovery technique owing to its fast adsorption speed, high loading capacity, low requirements on the clarity of solutions, simultaneous elution and regeneration at ambient temperature through the elaborate choice of eluent. In addition, ion-exchange resins can be custom-made to selectively extract gold because the functional groups can be designed to have high affinity for objective ions in the solution [74–80]. Therefore, resin adsorption is more suitable for gold recovery from pregnant thiosulfate solutions.

In this review, recent advances of gold recovery by the resin adsorption technique from pregnant thiosulfate solutions will be presented in detail. The advantages of strong-base resins over weak-base resins in gold adsorption and the differences of three kinds of elution principles of gold-loaded resins are indicated. Afterwards, a relatively comprehensive summary of the existing problems of the resin adsorption technique will be outlined. In the end, the potential development direction will also be proposed to solve these problems.

2. Progress of Gold Recovery by the Resin Adsorption Technique

Ion-exchange resin can be used to recover gold from pregnant thiosulfate solutions through the reversible ion-exchange reactions between the counter ions in the resins and gold (I) thiosulfate complex in the solutions. Gold exists predominantly in the form of $[Au(S_2O_3)_2]^{3-}$ anion in the solutions, and thus the resins used in gold recovery are all anion exchange resins, including strong and weak-base resins.

2.1. The Adsorption of Gold on the Ion-Exchange Resins

2.1.1. The Adsorption of Gold on the Weak-Base Resins

Weak-base resins have primary, secondary or tertiary amine functional groups (or a mixture of them) and their ion-exchange properties are dominated by the solution pH values. In the free-base form, the resins are unable to adsorb the gold (I) thiosulfate complex and thus they must be protonated usually by adding an acid prior to their use for gold recovery Equation (1) [75]. Generally, the greater the number of the protonated amine groups on the resin at a specific pH value, the higher the potential for a high gold loading capacity. After protonation, the resin can be used to extract gold from pregnant thiosulfate solutions Equation (1).

$$|-NR_2 + HX \Leftrightarrow |-NR_2H^+X^- \quad (1)$$

$$3|-NR_2H^+X^- + [Au(S_2O_3)_2]^{3-} \Leftrightarrow (|-NR_2H)_3{}^+[Au(S_2O_3)_2]^{3-} + 3X^- \quad (2)$$

where, the symbol $|-$ indicates the inert backbone, R denotes the amine functional group and X represents the counter ions that can be exchanged such as chloride or sulfate.

For weak-base resins, the specific pH needed for protonation is defined by pKa, i.e., the pH value that 50% of the functional groups are protonated. Typical pKa is in the range of 6–8, and thus most of the weak-base resins will not be protonated adequately in the range of 9–11, which is the common pH range of thiosulfate solutions [76]. Also, gold loading on weak-base resins markedly decreases with the increase of pH from 8 to 11 [77]. So, the gold loading ability of weak-base resins is usually very low when they are used to adsorb gold from pregnant thiosulfate solutions, and this has been demonstrated by certain research [77,78]. In addition, although the gold (I) thiosulfate complex is the only desired anion for adsorption, a considerable number of unwanted anions, including copper (I) thiosulfate complexes, sulfur-oxygen anions and other metal thiosulfate complexes in the solutions, can be adsorbed causing the dramatic decrease of gold loading and thus the significant increase of the resin dosage. Therefore, the use of weak-base resins to recover gold from pregnant thiosulfate solutions is not an efficient and economical choice.

2.1.2. The Adsorption of Gold on the Strong-Base Resins

Unlike weak-base resins, strong-base resins contain ammonium functional groups and are not limited by the protonation to adsorb the objective ions from solutions [75]. That is, strong-base resins are applicable to gold recovery over a broad pH range and their gold loading capacity is essentially independent of the solution pH values [77]. Therefore, they can be directly used to recover gold from pregnant thiosulfate solutions, as portrayed in Equation (3).

$$|-NR_3{}^+X^- + [Au(S_2O_3)_2]^{3-} \Leftrightarrow (|-NR_3)_3{}^+[Au(S_2O_3)_2]^{3-} + 3X^- \quad (3)$$

Strong-base resins are generally preferred over weak-base resins in terms of gold loading capacity under the given experimental conditions. The test results for these two kinds of resins indicated that gold loading on various strong-base resins reached up to 10-25 kg/t, but only less than 2 kg/t gold loading capacity was attained for weak-base resins [77]. Since strong-base resins have higher gold loading capacity, the effects of the competitive adsorption of other undesirable anions are generally acceptable. Therefore, strong-base resins are a preferred option for the gold recovery from pregnant thiosulfate solutions.

However, strong-base resins also readily adsorb other undesirable anions and consequently show poor selectivity for gold (I) thiosulfate complexes. Evidently, the presence of these competitive anions will decrease gold loading capacity of the resins remarkably. The effects of copper (I) thiosulfate complexes and tetrathionate on the gold loading on various strong-base resins has been studied by Zhang and Dreisinger [77,79]. The gold loading was closely related to the copper concentration in solution and the maximum gold loading dropped by about 70% as the copper concentration increased from 100 to 500 ppm. Tetrathionate was also significantly detrimental to gold adsorption. With the addition of 0.01 M tetrathionate, gold loading on various resins decreased by nearly 90%. However, the sustained increase of tetrathionate concentration did not make further difference because most of the active sites of the resins were occupied by tetrathionate, i.e., the resins were poisoned.

In addition to tetrathionate and copper (I) thiosulfate complex, other sulfur-oxygen anions and metal thiosulfate complexes, such as $S_3O_6^{2-}$, SO_3^{2-}, $[Ag(S_2O_3)_2]^{3-}$, $[Pb(S_2O_3)_2]^{2-}$ and $[Zn(S_2O_3)_2]^{2-}$, are also present in the real leach solutions. These anions can also be adsorbed on the strong-base resins and thus exert an important influence on gold adsorption. A detailed investigation has been conducted with respect to the effects of common sulfur-oxygen anions and metal thiosulfate complexes on the equilibrium loading of gold on the Amberjet 4200 strong-base resin [80]. It was proposed that the affinity orders of the resin for these two kinds of competitive anions in the solution were $[Au(S_2O_3)_2]^{3-} > S_3O_6^{2-}, S_4O_6^{2-} > SO_3^{2-} > S_2O_3^{2-} > SO_4^{2-}$ and $[Au(S_2O_3)_2]^{3-} > [Pb(S_2O_3)_2]^{2-} >> [Ag(S_2O_3)_2]^{3-} > [Cu(S_2O_3)_3]^{5-} >> [Zn(S_2O_3)_2]^{2-}$, respectively. The results indicate that trithionate, tetrathionate and lead (II) thiosulfate complex may have great impact on the equilibrium loading of gold. In addition, although the affinity of the resin for the $[Cu(S_2O_3)_3]^{5-}$ anion is weaker than that for the $[Au(S_2O_3)_2]^{3-}$ anion, the concentration of $[Cu(S_2O_3)_3]^{5-}$ anion is much higher than that of $[Au(S_2O_3)_2]^{3-}$ anion in a typical leach solution. As a result of this, the simultaneous loading of a considerable amount of copper with gold on the resins has become one of the primary factors affecting the gold recovery from pregnant thiosulfate solutions.

In order to solve the competitive adsorption problem of copper with gold on the resins, the replacement of traditional cupric-ammonia catalysis has been proposed [81]. Dowex 21K resin was adopted to recover gold from pregnant thiosulfate solutions with nickel catalysis. Nickel was not adsorbed on the resin in the range of 0.0005–0.05 mol/L Ni^{2+} and 0.05–0.2 mol/L $S_2O_3^{2-}$, and the maximum gold loading capacity of the resin achieved 95 kg/t, which is notably higher than that of common strong-base resins in pregnant thiosulfate solutions with copper catalysis. Also, a comparative study about leaching and recovery of gold using ammoniacal thiosulfate solutions with copper, nickel and cobalt catalysis has been conducted in the authors' laboratory [82]. The thiosulfate consumption of copper catalysis is much higher than that of nickel and cobalt catalysis. The competitive adsorption of nickel or cobalt along with gold did not occur when Tulsion A-21S strong-base resin was used to adsorb gold from their pregnant thiosulfate solutions. Evidently, the absence of competitive adsorption of nickel or cobalt with gold is extremely advantageous to the gold recovery because gold loading on the resin will increase substantially and the problem in separating gold and nickel or cobalt on the resin will also not need to be considered, therefore reducing the gold recovery cost considerably.

2.2. The Elution of Gold-Loaded Resins

The selection of eluents is critical because it not only determines the feasibility of cyclic utilization of leach solutions to some degree but also affects the subsequent gold recovery from the eluate solution.

In addition, elution process of gold-loaded resins is complex due to the loading of a large quantity of copper on the resins, and thus the gold and copper can be considered to be eluted simultaneously or separately. For one-step elution process, additional operations for separating these two metals from the eluate solution were needed in order to obtain final high-purity gold product, but this can easily cause the gold loss. Therefore, the complete separation of gold and copper from the resin is a preferred choice through the elaborate choice of eluents for them. Since the affinity of the common gold extraction resins for gold (I) thiosulfate complex is higher than that for copper (I) thiosulfate complex, copper will also be stripped during gold elution. Thus, pre-elution of copper is desirable to prevent the final gold product from being contaminating by the copper. The copper can be selectively pre-eluted with the solutions of oxygenated ammonia, ammonia-ammonium sulfate, ammonium thiosulfate, etc. and hence this paper mainly concentrates on the elution of gold based on different elution principles.

2.2.1. The Elution of Gold by Chemical Reaction

As shown in Figure 2(①), the resin is loaded with large amounts of $[Au(S_2O_3)_2]^{3-}$ anions when adsorption is completed. The chemical reaction principle is that the original $[Au(S_2O_3)_2]^{3-}$ anions are transformed to cationic gold complexes through the replacement of thiosulfate ligand [80]. The resins have noticeably reduced affinity for the new-formed cationic complex than gold (I) thiosulfate complex, and thus gold can be readily eluted from the resins. Unfortunately, few eluents can be applied to strip gold on the resins by this principle because the alternative ligand that can replace thiosulfate and complex with gold (I) to form cationic complex is scarce. Acidic thiourea is the only reported for the elution of gold (I) thiosulfate complex by chemical reaction principle [45,83], and the elution reaction of gold is illustrated in Equation (4).

$$(|-NR_3)_3{}^+[Au(S_2O_3)_2]^{3-} + 2SC(NH_2)_2 + 4HCl \\ \rightarrow 3|-NR_3{}^+Cl^- + Au[SC(NH_2)_2]_2{}^+Cl^- + 2S\downarrow + 2SO_2\uparrow + 2H_2O \quad (4)$$

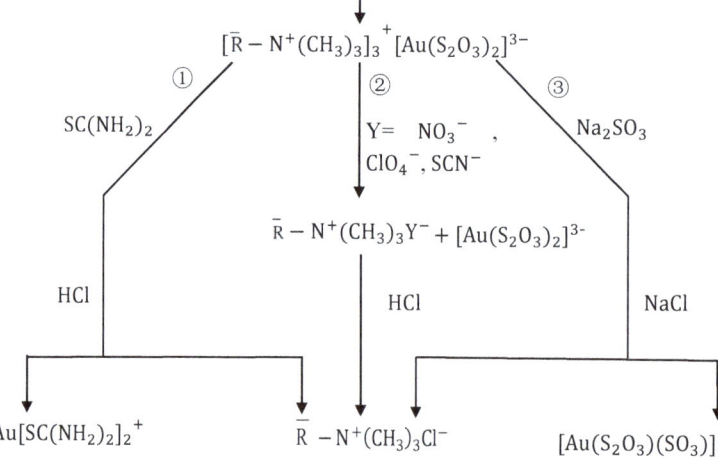

Figure 2. The elution of gold from strong-base resins (①, ② and ③ represent the elution procedure of $[Au(S_2O_3)_2]^{3-}$ anions loaded on the resin through the chemical reaction, displacement and synergistic ion exchange principles, respectively).

For this gold elution process, regeneration step of the resins is not necessary because the resins only have weak affinity for the chloride, and it can be readily exchanged by $[Au(S_2O_3)_2]^{3-}$ anions when the resins are used in the next adsorption circuit. Unfortunately, acidic thiourea is not a viable option because the sulfur-oxygen anions loaded on the resins such as thiosulfate and polythionates are not stable under acidic condition, and sulfur precipitation driving from the breakdown of them can poison the resin. Furthermore, osmotic shock produced by repeated elution in acidic media and adsorption in alkaline media can destroy the resin structure and thus increase recovery cost.

2.2.2. The Elution of Gold by Displacement

As indicated in Figure 2(②), the displacement principle uses an anion with high affinity for the resin to exchange adsorbed $[Au(S_2O_3)_2]^{3-}$ anions [80], which is also the most common method to be used in the gold elution. From the perspective of adsorption, the method is essentially achieved by altering the equilibrium of Equation (3) to the left through the increase of the concentration of X^- anion. On the basis of this principle, some eluents including thiocyanate, tetrathionate, nitrate and perchlorate have been founded to strip the gold effectively. However, gold elution by this principle usually requires high concentration of eluent because of the strong affinity of the resin for $[Au(S_2O_3)_2]^{3-}$ anion, therefore resulting in relatively high reagent cost. Furthermore, it is necessary to adopt an additional regeneration step to change the resin into its original form after gold elution. Otherwise, gold loading capacity of the resin will decrease dramatically due to the sustained accumulation of eluent anions on the resin.

Thiocyanate has been considered to be capable of effectively eluting loaded gold cyanide complex from anion exchange resins because of its strong affinity for the resin [84]. It was also used to elute gold (I) thiosulfate complex from the resin by the same authors, as described in Equation (5). After elution, the gold in the eluate solution was then recovered by conventional electrowinning [85]. However, thiocyanate is not an economical and environmental eluent, and the electrodes can be corroded during the subsequent gold electrowinning from the thiocyanate eluates [84]. After gold elution, sulfuric or hydrochloric acids can be used to regenerate the resin, but the thiocyanate ion degrades readily to elemental sulfur under strongly acidic condition. This is regarded to be undesirable from resin recycling point of view because sulfur precipitation will accumulate in the resin matrix and poison the resin by blocking resin pore. An alternative is to use ferric sulfate to remove the loaded thiocyanate and regenerate the resin to the sulfate form. Meanwhile, the eluted Fe (III) thiocyanate can be recovered by the addition of hydroxide to form Fe (III) hydroxide precipitate and the generated thiocyanate ions can be returned to elution cycle, as presented in Equations (6) and (7). However, this resin regeneration process may lead to resin breakage from osmotic shock driving from the change in pH from elution to regeneration.

$$(|-NR_3)_3{}^+[Au(S_2O_3)_2]^{3-} + 3SCN^- \Leftrightarrow 3|-NR_3{}^+SCN^- + [Au(S_2O_3)_2]^{3-} \quad (5)$$

$$4|-NR_3{}^+SCN^- + Fe_2(SO_4)_3 \rightarrow 2(|-NR_3)_2{}^+SO_4{}^{2-} + 2Fe(SCN)_2{}^+ + SO_4{}^{2-} \quad (6)$$

$$Fe(SCN)_2{}^+ + 3OH^- \rightarrow Fe(OH)_3 \downarrow + 2SCN^- \quad (7)$$

As a common anion in the pregnant thiosulfate solutions, $S_4O_6{}^{2-}$ anion can strongly compete with $[Au(S_2O_3)_2]^{3-}$ anion for the active sites of the resin, as indicated in Section 2.1.2. As a result of this, it has also been used to effectively elute gold and the stripped resin was regenerated with sulfide ions to convert tetrathionate to thiosulfate [86]. The reactions of gold (I) thiosulfate elution and resin regeneration can be portrayed in Equations (8) and (9).

$$2(|-NR_3)_3{}^+[Au(S_2O_3)_2]^{3-} + 3S_4O_6{}^{2-} \Leftrightarrow 3(|-NR_3)_2{}^+S_4O_6{}^{2-} + 2[Au(S_2O_3)_2]^{3-} \quad (8)$$

$$(|-NR_3)_2{}^+S_4O_6{}^{2-} + 2S^{2-} \Leftrightarrow (|-NR_3)_2{}^+S^{2-} + 2S_2O_3{}^{2-} + S\downarrow \quad (9)$$

However, this elution process is also not economical because tetrathionate is highly unstable and will decompose easily, with the result that large amount of eluent solution will be needed. Therefore, the eluent must be freshly made each time the elution of the resin is carried out [80]. The formation of sulfur precipitation is also inevitable during the resin regeneration with sulfide ions, as shown in Equation 9. Furthermore, sulfide precipitates of gold, copper or other metals may form when the resins with sulfide counter ions are return to the gold adsorption stage, which is highly unbeneficial to both the gold recovery and cycle use of the resin.

Apart from thiocyanate and tetrathionate, nitrate could also be adopted to elute gold (Equation 10) and electrowinning was utilized to recover gold from the eluate solution [87]. However, compared with thiocyanate and tetrathionate, much more concentrated nitrate solutions are needed to achieve almost complete gold elution. The potential reason is that the affinity of the resins for nitrate is substantially lower than that of thiocyanate and tetrathionate. It has also been observed that the eluate solution containing gold (I) thiosulfate complex was not stable and black gold precipitate occurred during electrowinning. The potential reason is that gold (I) thiosulfate complex will dissociate due to the lack of the free thiosulfate in the eluate solution, which is further accelerated due to the oxidation decomposition of free thiosulfate. Finally, the gold (I) reacts with the sulfide ions in the solution resulting in the formation of black gold sulfide precipitate.

$$(|-NR_3)_3{}^+[Au(S_2O_3)_2]^{3-} + 3NO_3{}^- \Leftrightarrow 3|-NR_3{}^+NO_3{}^- + [Au(S_2O_3)_2]^{3-} \tag{10}$$

In a recent study on the gold elution from the same resin with different gold loading capacities, perchlorate has been found to be an effective eluent (Equation 11) [81]. For 2.0 g resin with 5 kg/t gold loading capacity, nearly 100% gold can be successfully eluted by 2.5 mol/L perchlorate and an eluate with average gold concentration of 300 mg/L was produced. Moreover, the feasibility of this elution process has been demonstrated by a detailed comparison with the conventional Zadra process that is utilized to elute the loaded $[Au(CN)_2]^-$ anions from the activated carbon (Table 2) [88]. It is clear that the perchlorate elution process is superior to the Zadra process with respect to elution time, temperature, gold concentration in the eluate solution, gold recovery, etc. In addition, successful gold recovery by electrowinning from the eluate can be realized because perchlorate is a common electrolyte in the electrochemistry study. Unfortunately, both high concentration of perchlorate solution and resin regeneration are also required.

$$(|-NR_3)_3{}^+[Au(S_2O_3)_2]^{3-} + 3ClO_4{}^- \Leftrightarrow 3|-NR_3{}^+ClO_4{}^- + [Au(S_2O_3)_2]^{3-} \tag{11}$$

Table 2. Comparison of $[Au(S_2O_3)_2]^{3-}$ anion elution from the resin by the perchlorate process and $[Au(CN)_2]^-$ anion elution from activated carbon by the Zadra process [81].

Parameters	Perchlorate Process			Zadra Process
Gold loading capacity of resin/carbon/(kg/t)	5	10	20	4–5
Elution time/h	4	4	4	30–48
Temperature/K	Ambient	Ambient	Ambient	363–373
Pressure/kPa	Atmosphere	Atmosphere	Atmosphere	400–500
Flow rate/(bed volume number/h)	3	6	12	1–2
Maximum gold concentration/(mg/L)	1250	1400	1700	1000
Average gold concentration/(mg/L)	333	400	400	150–400
Gold recovery	>99.5%	>99.5%	>99.5%	96–98%
Gold loading capacity of stripped resin/carbon/(kg/t)	<0.05	<0.05	<0.05	0.15

Note: Eluent of perchlorate elution process: 2.5 mol/L NaClO$_4$, eluent of Zadra process: 10 g/L NaOH + 2 g/L NaCN.

2.2.3. The Elution of Gold by Synergistic Ion Exchange

As presented in Figure 2(③), the third elution principle is based on the concept of "synergistic ion exchange". In this principle, the eluent is a mixture of a weak eluent and sulfite. It is hard for the weak eluent to effectively elute gold loaded on the resins even when high concentration of it is used due to the weak affinity of the resin for the eluent anion. However, with the addition of a small amount of sulfite, $[Au(S_2O_3)_2]^{3-}$ complex will be converted to $[Au(S_2O_3)(SO_3)]^{3-}$ complex [89]. The resin has very weak affinity for the new-formed mixed ligand gold (I) thiosulfate-sulfite complex, and thus it can be readily eluted by the weak eluent with relatively low concentration, which is favorable to the decrease of elution costs. Furthermore, resin regeneration is not needed because the weak eluent anion will be readily exchanged by $[Au(S_2O_3)_2]^{3-}$ anions when the resin is returned to adsorption circuit, therefore simplifying the whole gold recovery process greatly.

Based on the elution principle, a study about the elution of the gold-loaded resins was conducted [90]. After copper elution, the resin was then stripped to recover gold with a mixed eluent consisting of 2 M sodium chloride and 0.1 M sodium sulfite. Subsequently, electrowinning was utilized to recover gold from the eluate solution. Gold could hardly be eluted using single 2 M sodium chloride solutions, which was also demonstrated by the research of Mohansingh [47]. With the addition of 0.1 M sodium sulfite, however, a complete gold elution could be achieved within 12 BV (bed volume, i.e., the volume of the reactor that is occupied by ion exchange resin) under the eluent flowrate of 5 BV/h. When the eluent flowrate was decreased to 2 BV/h, elution was complete only within 8 BV. Therefore, the sulfite significantly facilitates the elution efficiency of sodium chloride, and the probable elution reaction is shown in Equation 12. Another important advantage of the sulfite is that it can reduce the formation of gold sulfide pricipitate during electrowinning and thus significantly improve gold recovery from the eluate solution. Moreover, the addition of sulfite to the eluent can convert higher polythionates including tetrathionate, pentathionate and hexathionate to thiosulfate and trithionate (Equations 13–15). However, it should also be noted that trithionate, the decomposition product of these higher polythionates, can still accumulate on the resin and eventually has a negative effect on gold loading in the adsorption stage, as indicated in Section 2.1.2. In addition, poisonous chlorine may produce during electrowinning due to the presence of high concentration of chloride and thus worsen working conditions.

$$(|-NR_3)_3^+[Au(S_2O_3)_2]^{3-} + SO_3^{2-} + 3Cl^- \Leftrightarrow 3|-NR_3^+Cl^- + [Au(S_2O_3)(SO_3)]^{3-} + S_2O_3^{2-} \tag{12}$$

$$S_4O_6^{2-} + SO_3^{2-} \Leftrightarrow S_2O_3^{2-} + S_3O_6^{2-} \tag{13}$$

$$S_5O_6^{2-} + 2SO_3^{2-} \Leftrightarrow 2S_2O_3^{2-} + S_3O_6^{2-} \tag{14}$$

$$S_6O_6^{2-} + 3SO_3^{2-} \Leftrightarrow 3S_2O_3^{2-} + S_3O_6^{2-} \tag{15}$$

3. The Main Limitations of the Resin Adsorption Technique

The resin adsorption technique is the most suitable choice to recover gold from pregnant thiosulfate solutions compared with other recovery techniques, as mentioned in Introduction. However, there are still certain problems to be addressed to further develop the technique.

3.1. The Competitive Adsorption of Undesirable Anions

Thiosulfate is generally considered as being metastable, which means that some oxidative decomposition reactions can occur in a typical leaching environment, as described in Equations (16) and (17). In addition, in the thiosulfate leaching system, copper and ammonia are added to accelerate gold dissolution through the formation of copper (II) ammine complexes. However, a problem associated with the use of cupric ammine is that it accelerates the oxidation of thiosulfate to produce tetrathionate, which is also unstable and can be easily degraded into trithionate and thiosulfate [91–95], as presented in Equations (18) and (19).

$$3S_2O_3^{2-} + 2O_2 + H_2O \rightarrow 2S_3O_6^{2-} + 2OH^- \tag{16}$$

$$4S_2O_3^{2-} + O_2 + 2H_2O \rightarrow 2S_4O_6^{2-} + 4OH^- \tag{17}$$

$$2[Cu(NH_3)_4]^{2+} + 8S_2O_3^{2-} \rightarrow 2[Cu(S_2O_3)_3]^{5-} + 8NH_3 + S_4O_6^{2-} \tag{18}$$

$$4S_4O_6^{2-} + 6OH^- \rightarrow 5S_2O_3^{2-} + 2S_3O_6^{2-} + 3H_2O \tag{19}$$

The degradation products of thiosulfate, particularly polythionates including trithionate and tetrathionate, will intensely compete with gold (I) thiosulfate complex for active sites in the resins, and low concentration of them in the leach solution will decrease gold loading capacity of the resins tremendously. Therefore, it is very necessary to minimize the generation of them in the leaching stage to ensure an efficient gold recovery from the pregnant thiosulfate solutions by the resin adsorption technique.

It should be noticed that the cupric ammine itself is also reduced to $[Cu(S_2O_3)_3]^{5-}$ when it accelerate the oxidation of thiosulfate, as shown in Equation 18. Copper (I) thiosulfate complex is more stable than its ammine complex thermodynamically [81], and thus copper (I) exists primarily as thiosulfate complex in the pregnant thiosulfate solutions. Substantial amounts of $[Cu(S_2O_3)_3]^{5-}$ can readily load on the resins with $[Au(S_2O_3)_2]^{3-}$ due to the same ligand of these two complex anions. For this reason, it is very difficult to selectively adsorb gold (I) thiosulfate complex from pregnant thiosulfate solutions. The co-adsorption of copper (I) also makes it necessary to adopt a complex and costly two-stage elution process to separate the copper and gold loaded on the resins.

Apart from polythionates and copper (I) thiosulfate complex, other metal thiosulfate complexes, such as $[Ag(S_2O_3)_2]^{3-}$, $[Pb(S_2O_3)_2]^{2-}$ and $[Zn(S_2O_3)_2]^{2-}$, may also exist in the real leach solutions [80]. Their effects on the gold adsorption on the resins can also usually not be neglected because they are able to load on the resins. The competitive adsorption of them with gold (I) thiosulfate complex not only decreases the gold loading capacity of the resins but also complicates subsequent elution process.

3.2. The Lack of Suitable Gold Extraction Resin and Eluent

Most of the available resins used in current research on gold recovery from pregnant thiosulfate solutions are those which show good performance for gold recovery from pregnant cyanide solutions (Table 3) [96–99]. However, these resins are generally not highly selective for $[Au(S_2O_3)_2]^{3-}$ anion because they were designed to adsorb singly charged $[Au(CN)_2]^-$ anion. Furthermore, the presence of a series of competitive anions, including copper (I) thiosulfate complexes, sulfur-oxygen anions and other metal thiosulfate complexes in the solutions also places greater demands on the selectivity of the resin. Nevertheless, the selectivity of the resin is a complex subject, which is mainly determined by the properties of the resin itself and the anion. For the resin, the functional group, hydrophobicity of the polymeric matrix, porosity and ion identity (number of ionic groups per unit volume) of the resin are all important factors influencing its selectivity for the anion [100,101]. Also, previous studies on the adsorption of metal cyanide anions on both strong- and weak-base resins indicated that the properties of the anion, including polarisability, charge density (the ratio of the charge to the number of atoms), the degree of hydration and the shape or size, give an indication of what type of resin will be suitable for gold recovery [101–104]. Taking the above properties of the anion into consideration in a thiosulfate solution, O'Malley [80] proposed the likely selectivity orders of the resin for sulfur-oxygen anions and metal thiosulfate complexes (Tables 4 and 5), which also were supported by certain research results [105–107]. However, research on the selectivity of the resin for $[Au(S_2O_3)_2]^{3-}$ anion has been rare up to now and there is still much work to be done to design a resin highly selective for gold (I) thiosulfate complex.

Table 3. The resins used for gold recovery from pregnant thiosulfate solutions.

Resin	Type	[Au][1]	[Cu][1]	[$S_2O_3^{2-}$]	[NH_3]	pH	T/°C	Time/h	Au/%[2]	Reference
AV-17-10P	R_4N^+	9.5–17.9	-	0.5	0.5	10.8–11.0	-	5	94.2	[45]
Amberlite IRA 400	R_4N^+	20	196.85	0.18	0.324	10	25	0.33	100	[75]
Purolite A500C	R_4N^+	1.8	22	0.05	0.1	8	60	6	99.45	[96]
Amberlite IRA 400	R_4N^+	9.27	125	1.0	0.1	9	20	10	94.7	[47]
Amberjet 4200	R_4N^+	10	10	0.5	0.2	9.5	-	-	99	[80]
Dowex 21K	R_4N^+	20	500	0.1	0.2	11	23–25	24	100	[77]
Dowex 21K	R_4N^+	100	29.5[3]	0.05	0.5	9.5	-	3	95	[81]
Amberlite IRA 410	R_4N^+	8	100	0.1	0.1	11	25	2	>90	[97]
Purolite A530	R_4N^+	39.4	-	0.5	-	10	-	-	94	[78]
Dowex G51	R_4N^+	10	500	0.1	-	11.7	23	-	98	[98]
Amberlite IRA-93	WB	10	-	0.1	-	8	23	2	94.3	[98]
Aurix 100	Guan	1–8	-	0.00674–0.04	0.30–0.81	9–10.5	25–40	0–3	99	[99]

Note: All the results were obtained under batch experiment condition. R_4N^+: quaternary ammonium (strong base); WB: weak base; Guan: guanidyl. "-" denotes "absence" of the items or "the literature did not mention", "T" represents temperature. [1] Reported in ppm. [2] Gold recovery (%). [3] Nickel concentration.

Table 4. Predicted selectivity of the resin for sulfur-oxygen anions presented in declining order.

Name	Formula	Formula Weight	Polarizability	Charge Density	Hydration
Pentathionate	$S_5O_6^{2-}$	256.5	High	0.18	High
Tetrathionate	$S_4O_6^{2-}$	224.3	High	0.2	High
Trithionate	$S_3O_6^{2-}$	192.2	High	0.22	High
Dithionate	$S_2O_6^{2-}$	160.1	Medium	0.25	High
Dithionite	$S_2O_4^{2-}$	128.2	Medium	0.33	Medium
Thiosulfate	$S_2O_3^{2-}$	112.2	Medium	0.4	Medium
Sulfate	SO_4^{2-}	96.1	Low	0.4	Medium
Sulfite	SO_3^{2-}	80.1	Low	0.5	Medium

Table 5. Predicted selectivity of the resin for metal thiosulfate complexes presented in declining order.

Name	Formula	Formula Weight	Polarizability	Charge Density	Hydration
Silver (I) thiosulfate	$[Ag(S_2O_3)_3]^{5-}$	444.3	High	0.13	NK
Copper (I) thiosulfate	$[Cu(S_2O_3)_3]^{5-}$	400.1	High	0.13	NK
Lead (II) thiosulfate	$[Pb(S_2O_3)_3]^{4-}$	543.7	High	0.25	NK
Gold (I) thiosulfate	$[Au(S_2O_3)_2]^{3-}$	421.3	High	0.27	Low
Lead (II) thiosulfate	$[Pb(S_2O_3)_2]^{2-}$	431.5	High	0.18	NK
Silver (I) thiosulfate	$[Ag(S_2O_3)_2]^{3-}$	332.2	High	0.27	Low
Copper (I) thiosulfate	$[Cu(S_2O_3)_2]^{3-}$	287.8	High	0.27	NK
Zinc (II) thiosulfate	$[Zn(S_2O_3)_2]^{2-}$	289.6	High	0.18	High

Note: "NK" denotes "not known".

Gold can be effectively adsorbed by strong-base resins from pregnant thiosulfate solutions. The gold-loaded resins are pre-eluted to remove copper, and the pre-eluate containing high concentration of copper will be returned to the leaching circuit. However, to seek for an applicable eluent of gold has been proven problematic mainly owning to the strong affinity of the resins for $[Au(S_2O_3)_2]^{3-}$ anion. Most of the eluents proposed previously, such as the single component solutions of thiocyanate, tetrathionate, nitrate and perchlorate or the two component solutions such as thiourea + sulfuric/hydrochloric acids and sodium sulfite + sodium chloride, are all not ideal due to either high reagent cost or environmental pollution, and the characteristics of them are summarized and evaluated in Table 6. Therefore, the development of an appropriate eluent to elute the gold from the resins will be one of the feasible measures to further develop the resin adsorption technique.

Table 6. Evaluation of various elution processes for $[Au(S_2O_3)_2]^{3-}$ anion.

Elution Process		Characteristics	References
Single component	Thiocyanate	• Strong affinity for resin and fast elution kinetics • Resin regeneration is needed and complex, and high resin losses may occur • High reagent cost and environmental concerns • Corrosion of electrodes in electrowinning cell	[85]
	Tetrathionate	• Strong affinity for resin and fast elution kinetics • High reagent consumption due to stability issues • Reasonably high temperature is required and the formation of sulfur precipitation is also inevitable for the resin regeneration with sulfide ions • Metal sulfide precipitates can form when the resins with sulfide counter ions are recycled back to the gold adsorption stage	[86]
	Nitrate	• Weak affinity for resin and slow elution kinetics • High concentration of eluent solution is needed • Black gold precipitate occurs during subsequent electrowinning	[80,87]
	Perchlorate	• Strong affinity for resin and fast elution kinetics • Successful electrowinning can be realized for the gold eluate because perchlorate is a common electrolyte in the study of electrochemistry • Resin regeneration and high concentration of eluent solution is also required	[81]
Two components	Thiourea + sulfuric/hydrochloric acids	• Resin does not require regeneration after elution • Sulfur precipitation can produce due to the decomposition of sulfur-oxygen anions under acidic condition • High resin losses may produce due to the osmotic shock produced by repeated elution in acidic media and adsorption in alkaline media	[45,83]
	Sodium chloride + sodium sulfite	• Reduced affinity of gold (I) complex for the resin by the formation of mixed ligand gold (I) thiosulfate-sulfite complex in the presence of sulfite • Lower concentrations of the eluent (sodium chloride) and more efficient elution of the gold can be realized • Trithionate can accumulate on the resin and poisonous chlorine may produce during electrowinning	[90]

4. Future Development

Current research about gold recovery by ion exchange resins from pregnant thiosulfate solutions is in its infancy and a large number of further work is needed to make the recovery technique more efficient, economical and environmentally friendly. In terms of the main problems existing in the resin adsorption technique, the following countermeasures can be taken to optimize the technique.

(1) A suitable pretreatment can be conducted to remove the base metals before leaching to reduce their detrimental effects on subsequent gold recovery. To weaken or eliminate the competitive adsorption effects of copper (I) thiosulfate complexes and polythionates, one feasible measure is to minimize their generation during leaching through the elaborate control of reaction conditions. Another more effective measure is the replacement of traditional cupric-ammonia catalysis with other metals such as nickel- and cobalt-based catalysts. It reduces the consumption of thiosulfate and thus decreases the formation of polythionates. Also, the competitive adsorption of nickel and cobalt with gold will not occur because they do not complex with thiosulfate to form stable Ni/Co-$S_2O_3^{2-}$ complexes, and therefore, the complicated and costly two-stage elution process can be substituted by a simple and low-cost one-stage process.

(2) The structure–activity relationship of the resin functional groups can be investigated through the first principle and quantum chemistry calculation to obtain the ideal resins that have strong affinity and high selectivity for gold (I) thiosulfate complex over copper (I) thiosulfate complex, polythionates and other metal thiosulfate complexes. As for the eluent, the mixed eluent of chloride and sulfite deriving from the concept of synergistic ion exchange needs to be further developed through the displacement of chloride by other environment-friendly weak eluents which have no negative effect on the subsequent electrowinning of gold eluate solutions.

Acknowledgments: Financial supports from the National Natural Science Foundation of China (grant No. 51504293 and 51574284), the China Postdoctoral Science Foundation (grant No. 2014M550422), Hunan Provincial Natural Science Foundation of China (grant No. 2015JJ3149), the Fundamental Research Funds for the Central Universities of Central South University (grant No. 2016zzts473), ,and the Open-End Fund for the Valuable and Precision Instruments of Central South University (CSUZC201704) are all gratefully acknowledged.

Author Contributions: Zhonglin Dong completed the main part of this review; Tao Jiang and Yongbin Yang completed the other parts of this review; Bin Xu and Qian Li offered advice for writing and revision of this review.

Conflicts of Interest: The authors declare no conflict of interest.

References

1. Jiang, T. *Chemistry of Extractive Metallurgy of Gold*; Hunan Science and Technology Press: Changsha, China, 1998.
2. Hilson, G.; Monhemius, A.J. Alternatives to cyanide in the gold mining industry: What prospects for the future? *J. Clean. Prod.* **2006**, *14*, 1158–1167. [CrossRef]
3. Xu, B.; Yang, Y.B.; Li, Q.; Jiang, T.; Liu, S.Q.; Li, G.H. The development of an environmentally friendly leaching process of a high C, As and Sb bearing sulfide gold concentrate. *Miner. Eng.* **2016**, *89*, 138–147. [CrossRef]
4. Braul, P. Thiosulfate going commercial. *Cim. Mag.* **2013**, *8*, 42–45.
5. Choi, Y.; Baron, J.Y.; Wang, Q.; Langhans, J.; Kondos, P. Thiosulfate processing—From lab curiosity to commercial application. In Proceedings of the World Gold 2013, Brisbane, QLD, Australia, 26–29 September 2013.
6. Xu, B.; Kong, W.H.; Li, Q.; Yang, Y.B.; Jiang, T. A review of thiosulfate leaching of gold: Focus on thiosulfate consumption and gold recovery from pregnant solution. *Metals* **2017**, *7*, 222. [CrossRef]
7. Baron, J.Y.; Mirza, J.; Nicol, E.A.; Smith, S.R.; Leitch, J.J.; Choi, Y.; Lipkowski, J. SERS and electrochemical studies of the gold–electrolyte interface under thiosulfate based leaching conditions. *Electrochim. Acta* **2013**, *111*, 390–399. [CrossRef]
8. Zhang, X.M.; Senanayake, G. A review of ammoniacal thiosulfate leaching of gold: An update useful for further research in non–cyanide gold lixiviants. *Miner. Process Extr. Metall. Rev.* **2016**, *37*, 385–411. [CrossRef]
9. Lampinen, M.; Laari, A.; Turunen, I. Ammoniacal thiosulfate leaching of pressure oxidized sulfide gold concentrate with low reagent consumption. *Hydrometallurgy* **2015**, *151*, 1–9. [CrossRef]
10. Senanayake, G. Gold leaching by copper (II) in ammoniacal thiosulphate solutions in the presence of additives. Part I: A review of the effect of hard–soft and Lewis acid–base properties and interactions of ions. *Hydrometallurgy* **2012**, *115–116*, 1–20. [CrossRef]
11. Liu, X.L.; Xu, B.; Min, X.; Li, Q.; Yang, Y.B.; Jiang, T.; He, Y.H.; Zhang, X. Effect of pyrite on thiosulfate leaching of gold and the role of ammonium alcohol polyvinyl phosphate (AAPP). *Metals* **2017**, *7*, 278. [CrossRef]
12. Xu, B.; Yang, Y.B.; Li, Q.; Jiang, T.; Li, G.H. Stage leaching of a complex polymetallic sulfide concentrate: Focus on the extraction of Ag and Au. *Hydrometallurgy* **2016**, *159*, 87–94. [CrossRef]
13. Xu, B.; Yang, Y.B.; Jiang, T.; Li, Q.; Zhang, X.; Wang, D. Improved thiosulfate leaching of a refractory gold concentrate calcine with additives. *Hydrometallurgy* **2015**, *152*, 214–222. [CrossRef]
14. Xu, B.; Yang, Y.B.; Li, Q.; Yin, W.; Jiang, T.; Li, G.H. Thiosulfate leaching of Au, Ag and Pd from a high Au, Ag and Pd bearing decopperized anode slime. *Hydrometallurgy* **2016**, *164*, 278–287. [CrossRef]
15. Yang, Y.B.; Zhang, X.; Xu, B.; Li, Q.; Jiang, T.; Wang, Y.X. Effect of arsenopyrite on thiosulfate leaching of gold. *Trans. Nonferrous Met. Soc. China* **2015**, *25*, 3454–3460. [CrossRef]
16. Feng, D.; Van Deventer, J.S.J. Thiosulphate leaching of gold in the presence of orthophosphate and polyphosphate. *Hydrometallurgy* **2011**, *106*, 38–45. [CrossRef]
17. Feng, D.; Van Deventer, J.S.J. Effect of thiosulphate salts on ammoniacal thiosulphate leaching of gold. *Hydrometallurgy* **2010**, *105*, 120–126. [CrossRef]
18. Feng, D.; Van Deventer, J.S.J. Thiosulphate leaching of gold in the presence of ethylenediaminetetraacetic acid (EDTA). *Miner. Eng.* **2010**, *23*, 143–150. [CrossRef]
19. Feng, D.; Van Deventer, J.S.J. The role of amino acids in the thiosulphate leaching of gold. *Miner. Eng.* **2014**, *24*, 1022–1024. [CrossRef]
20. Liu, X.L.; Xu, B.; Yang, Y.B.; Li, Q.; Jiang, T.; Zhang, X.; Zhang, Y. Effect of galena on thiosulfate leaching of gold. *Hydrometallurgy* **2017**, *17*, 157–164. [CrossRef]
21. Xu, B.; Yang, Y.B.; Li, Q.; Jiang, T.; Zhang, X.; Li, G.H. Effect of common associated sulfide minerals on thiosulfate leaching of gold and the role of humic acid additive. *Hydrometallurgy* **2017**, *17*, 44–52. [CrossRef]
22. Jiang, T.; Xu, S.; Chen, J. Gold and silver extraction by ammoniacal thiosulfate catalytical leaching at ambient temperature. In Proceedings of the First International Conference of Modern Process Mineralogy and Mineral Processing, Beijing, China, 22–25 September 1992.

23. Jiang, T.; Chen, J.; Xu, S. Electrochemistry and mechanism of leaching gold with ammoniacal thiosulfate. In Proceedings of the XVIII International Mineral Processing Congress, Sydney, Australia, 23–28 May 1993.
24. Breuer, P.L.; Jeffrey, M.I. An electrochemical study of gold leaching in thiosulfate solutions containing copper and ammonia. *Hydrometallurgy* **2002**, *65*, 145–157. [CrossRef]
25. Arima, H.; Fujita, T.; Yen, W.T. Using nickel as a catalyst in ammonium thiosulfate leaching for gold extraction. *Mater. Trans.* **2004**, *45*, 516–526. [CrossRef]
26. Aylmore, M.G. Treatment of a refractory gold-copper sulfide concentrate by copper ammoniacal thiosulfate leaching. *Miner. Eng.* **2001**, *14*, 615–637. [CrossRef]
27. Aylmore, M.G.; Muir, D.M.; Staunton, W.P. Effect of minerals on the stability of gold in copper ammoniacal thiosulfate solutions—The role of copper, silver and polythionates. *Hydrometallurgy* **2014**, *143*, 12–22. [CrossRef]
28. Wang, R.Y.; Brierley, J.A. Thiosulfate leaching follwing biooxidation pre-treatment for gold recovery from refractory carbonaceous-sulfidic ore. *Miner. Eng.* **1997**, *49*, 76–80.
29. Chandra, I.; Jeffrey, M.I. A fundamental study of ferric oxalate for dissolving gold in thiosulfate solutions. *Hydrometallurgy* **2005**, *77*, 191–201. [CrossRef]
30. Ficeriova, J.; Balaz, P.; Boldizarova, E.; Jelen, S. Thiosulfate leaching of gold from a mechanically activated CuPbZn concentrate. *Hydrometallurgy* **2002**, *67*, 37–43. [CrossRef]
31. Senanayake, G.; Zhang, X.M. Gold leaching by copper(II) in ammoniacal thiosulfate solutions in the presence of additives. Part II: Effect of residual Cu(II), pH and redox potentials on reactivity of colloidal gold. *Hydrometallurgy* **2012**, *115–116*, 21–29. [CrossRef]
32. Yu, H.; Zi, F.T.; Hu, X.Z.; Zhong, J.; Nie, Y.H.; Xiang, P.Z. The copper–ethanediamine–thiosulphate leaching of gold ore containing limonite with cetyltrimethyl ammonium bromide as the synergist. *Hydrometallurgy* **2014**, *150*, 178–183. [CrossRef]
33. Arima, H.; Fujita, T.; Yen, W.T. Gold Cementation from ammonium thiosulfate solution by zinc, copper and aluminium powders. *Mater. Trans.* **2002**, *43*, 485–493. [CrossRef]
34. Hu, J.X.; Gong, Q. Recovery of gold from thiosulfate solution. *Eng. Chem. Metall.* **1989**, *10*, 45–50. (In Chinese)
35. Guerra, E.; Dreisinger, D.B. A study of the factors affecting copper cementation of gold from ammoniacal thiosulphate solution. *Hydrometallurgy* **1999**, *51*, 155–172. [CrossRef]
36. Choo, W.L.; Jeffrey, M.I. An electrochemical study of copper cementation of gold(I) thiosulfate. *Hydrometallurgy* **2004**, *71*, 351–362. [CrossRef]
37. Hiskey, J.B.; Lee, J. Kinetics of gold cementation on copper in ammoniacal thiosulfate solution. *Hydrometallurgy* **2003**, *69*, 45–56. [CrossRef]
38. Lee, J. Gold Cementation on Copper in Thiosulfate Solution: Kinetic, Electrochemical, and Morphological Studies. Ph.D. Thesis, Department of Materials Science and Engineering, University of Arizona, Tucson, AZ, USA, December 2003.
39. Awadalla, F.T.; Ritcey, G.M. Recovery of gold from thiourea, thiocyanate or thiosulfate solutions by reduction–precipitation with a stabilized form of sodium borohydride. *Sep. Sci. Technol.* **1990**, *26*, 1207–1228. [CrossRef]
40. Groves, W.D.; Blackman, L. Recovery of Precious Metals from Evaporite Sediments. U.S. Patent 5,405,430, 11 April 1995.
41. Kerley, B.J. Recovery of Precious Metals from Difficult Ores. U.S. Patent 4,269,622, 26 May 1981.
42. Navarro, P.; Vargas, C.; Alonso, M.; Alguacil, F.J. The adsorption of gold on activated carbon from thiosulfate-ammoniacal solutions. *Gold Bull.* **2006**, *39*, 93–97. [CrossRef]
43. Navarro, P.; Vargas, C.; Alonso, M.; Alguacil, F.J. Towards a more environmentally friendly process for gold: Models on gold adsorption onto activated carbon from ammoniacal thiosulfate solutions. *Desalination* **2007**, *211*, 58–63. [CrossRef]
44. Marchbank, A.R.; Thomas, K.G.; Dreisinger, D.; Fleming, C. Gold Recovery from Refractory Carbonaceous Ores by Pressure Oxidation and Thiosulfate Leaching. U.S. Patent 5,536,297, 28 July 1996.
45. Kononova, O.N.; Kholmogorov, A.G.; Kononov, Y.S.; Pashkov, G.L.; Kachin, S.V.; Zotova, S.V. Sorption recovery of gold from thiosulphate solutions after leaching of products of chemical preparation of hard concentrates. *Hydrometallurgy* **2001**, *59*, 115–123. [CrossRef]
46. Vargas, C.; Navarro, P.; Araya, E.; Pavez, F.; Alguacil, F.J. Recovery of gold from solutions with ammonia and thiosulfate using activated carbon. *Rev. Metal.* **2006**, *42*, 222–233. [CrossRef]

47. Mohansingh, R. Adsorption of Gold from Gold Copper Ammonium Thiosulfate Complex onto Activated Carbon and Ion Exchange Resins. Master's Thesis, University of Nevada, Reno, NV, USA, May 2000.
48. Aylmore, M.G.; Muir, D.M. Thiosulfate leaching of gold—A review. *Miner. Eng.* **2001**, *14*, 135–174. [CrossRef]
49. Gallagher, N.P.; Hendrix, J.L.; Milosavljevic, E.B.; Nelson, J.H.; Solujic, L. Affinity of activated carbon towards some gold(I) complexes. *Hydrometallurgy* **1990**, *25*, 305–316. [CrossRef]
50. Gallagher, N.P. The affinity of carbon for gold complexes: dissolution of finely disseminated gold using a flow electrochemical cell. *J. Electrochem. Soc.* **1990**, *21*, 2546–2551. [CrossRef]
51. Lulham, J.P.; Lindsay, D. Gold Recovery from Thiosulfate Ore Leaching Solutions. International Patent WO/1991/011539, 8 August 1991.
52. Parker, G.K.; Gow, R.N.; Young, C.A.; Twidwell, L.G.; Hope, G.A. Spectroelectrochemical investigation of the reaction between adsorbed cuprous cyanide and gold thiosulfate ions at activated carbon surfaces. In Proceedings of the Hydrometallurgy 2008, Sixth International Symposium, Littleton, CO, USA, 17–20 August 2008.
53. Young, C.A.; Gow, R.N.; Twidwell, L.G.; Parker, G.K.; Hope, G.A. Cuprous cyanide adsorption on activated carbon: pretreatment for gold take-up from thiosulfate solutions. In Proceedings of the Hydrometallurgy 2008, Sixth International Symposium, Littleton, CO, USA, 17–20 August 2008.
54. Yu, H.; Zi, F.T.; Hu, X.Z.; Nie, Y.H.; Xiang, P.Z.; Xu, J.; Chi, H. Adsorption of the gold–thiosulfate complex ion onto cupric ferrocyanide(CuFC)-impregnated activated carbon in aqueous solutions. *Hydrometallurgy* **2015**, *154*, 111–117. [CrossRef]
55. Chen, J.Y.; Deng, T.; Zhu, G.C.; Zhao, J. Leaching and recovery of gold in thiosulfate based system—A research summary at ICM. *Trans. Indian Inst. Met.* **1996**, *49*, 841–849.
56. Zhao, J.; Wu, Z.C.; Chen, J.Y. Extraction of gold from thiosulfate solutions with alkyl phosphorus esters. *Hydrometallurgy* **1997**, *46*, 363–372. [CrossRef]
57. Zhao, J.; Wu, Z.C.; Chen, J.Y. Extraction of gold from thiosulfate solutions using amine mixed with neutral donor reagents. *Hydrometallurgy* **1998**, *48*, 133–144. [CrossRef]
58. Zhao, J.; Wu, Z.C.; Chen, J.Y. Gold extraction from thiosulfate solutions using mixed amines. *Solvent Extr. Ion Exch.* **1998**, *16*, 1407–1420. [CrossRef]
59. Zhao, J.; Wu, Z.C.; Chen, J.Y. Solvent extraction of gold in thiosulfate solutions with amines. *Solvent Extr. Ion Exch.* **1998**, *16*, 527–543. [CrossRef]
60. Zhao, J.; Wu, Z.C.; Chen, J.Y. Separation of gold from other metals in thiosulfate solutions by solvent extraction. *Sep. Sci. Technol.* **1999**, *34*, 2061–2068. [CrossRef]
61. Virnig, M.J.; Sierakoski, J.M. Ammonium Thiosulfate Complex of Gold or Silver and an Amine. U.S. Patent 6,197,214, 6 March 2001.
62. Liu, K.J.; Yen, W.T.; Shibayama, A.; Miyazaki, T.; Fujita, T. Gold extraction from thiosulfate solution using trioctylmethylammonium chloride. *Hydrometallurgy* **2004**, *71*, 41–53.
63. Grosse, A.C.; Dicinoski, G.W.; Shaw, M.J.; Haddad, P.R. Leaching and recovery of gold using ammoniacal thiosulfate leach liquors (a review). *Hydrometallurgy* **2003**, *69*, 1–21. [CrossRef]
64. Sullivan, A.M.; Kohl, P.A. Electrochemical study of the gold thiosulfate reduction. *J. Electrochem. Soc.* **1997**, *144*, 1686–1690. [CrossRef]
65. Sullivan, A.M.; Kohl, P.A. The autocatalytic deposition of gold in nonalkaline, gold thiosulfate electroless bath. *J. Electrochem. Soc.* **1995**, *142*, 2250–2255. [CrossRef]
66. Abbruzzese, C.; Fornari, P.; Massidda, R.; Veglio, F.; Ubaldini, S. Thiosulfate leaching for gold hydrometallurgy. *Hydrometallurgy* **1995**, *39*, 265–276. [CrossRef]
67. Osaka, T.; Kodera, A.; Misato, T.; Homma, T.; Okinaka, Y. Electrodeposition of soft gold from a thiosulfate–sulfite bath for electronics applications. *J. Electrochem. Soc.* **1997**, *144*, 3462–3469. [CrossRef]
68. Aledresse, A. Gold Recovery from Low Concentrations using Nanoporous Silica Adsorbent. Ph.D. Thesis, Laurentian University, Sudbury, ON, Canada, January 2009.
69. Boissiere, C.; Larbot, A.; Van der lee, A.; Kooyman, P.J.; Prouzet, E. A new synthesis of mesoporous MSU–X silica controlled by a two–step pathway. *Chem. Mater.* **2000**, *12*, 2902–2913. [CrossRef]
70. Fotoohi, B.; Mercier, L. Modification of pore structure and functionalization in MSU–X silica and application in adsorption of gold thiosulfate. *Microporous Mesoporous Mater.* **2014**, *190*, 255–266. [CrossRef]
71. Fotoohi, B.; Mercier, L. Recovery of precious metals from ammoniacal thiosulfate solutions by hybrid mesoporous silica: 2—A prospect of PGM adsorption. *Sep. Purif. Technol.* **2015**, *149*, 82–91. [CrossRef]

72. Fotoohi, B.; Mercier, L. Recovery of precious metals from ammoniacal thiosulfate solutions by hybrid mesoporous silica: 3—Effect of contaminants. *Sep. Purif. Technol.* **2015**, *139*, 4–24. [CrossRef]
73. Fotoohi, B.; Mercier, L. Recovery of precious metals from ammoniacal thiosulfate solutions by hybrid mesoporous silica: 1—Factors affecting gold adsorption. *Sep. Purif. Technol.* **2014**, *127*, 84–96. [CrossRef]
74. Grosse, A.C. The Development of Resin Sorbents Selective for Gold in Ammoniacal Thiosulfate Leach Liquors. Ph.D. Thesis, School of Chemistry, University of Tasmania, Australia, September 2006.
75. Atluri, V.P. Recovery of Gold and Silver from Ammoniacal Thiosulfate Solutions Containing Copper by Ion Exchange Resin Method. Master's Thesis, Department of Materials Science and Engineering, University of Arizona, Tucson, AZ, USA, December 1987.
76. Fleming, C.A.; Cromberge, G. The extraction of gold from cyanide solutions by strong–and weak–base anion-exchange resins. *J. South. Afr. Inst. Min. Metall.* **1984**, *8*, 125–137.
77. Zhang, H.G.; Dreisinger, D.B. The adsorption of gold and copper onto ion–exchange resins from ammoniacal thiosulfate solutions. *Hydrometallurgy* **2002**, *66*, 67–76. [CrossRef]
78. Kononova, O.N.; Shatnykh, K.A.; Prikhod'ko, K.V.; Kashirin, D.M. Ion exchange recovery of gold(I) and Silver (I) from thiosulfate solutions. *Russ. J. Phys. Chem. A* **2009**, *83*, 2340–2345. [CrossRef]
79. Zhang, H.G.; Dreisinger, D.B. The recovery of gold from ammonical thiosulfate solutions containing copper using ion exchange resin columns. *Hydrometallurgy* **2004**, *72*, 225–234. [CrossRef]
80. O'Malley, G.P. Recovery of Gold from Thiosulfate Solutions and Pulps with Anion Exchange Resins. Ph.D. Thesis, Murdoch University, Perth, Australia, March 2002.
81. Arima, H.; Fujita, T.; Yen, W.T. Gold recovery from nickel catalyzed ammonium thiosulfate solution by strongly basic anion exchange resin. *Mater. Trans.* **2003**, *44*, 2099–2107. [CrossRef]
82. Min, X. Research on the Electrochemistry of Thiosulfate Gold Leaching Catalyzed by Co/Ni-NH3. Master's Thesis, Central South University, Changsha, China, June 2017. (In Chinese)
83. Lai, C.S. The Analysis and Recovery of Gold from Ammonium Thiosulfate Leach Solutions. Master's Thesis, Kunming University of Science and Technology, Kunming, China, June 2011. (In Chinese)
84. Fleming, C.A.; Cromberge, G. The elution of aurocyanide from strong- and weak-base resins. *J. South. Afr. Inst. Min. Metall.* **1984**, *84*, 269–280.
85. Thomas, K.G.; Fleming, C.A.; Marchbank, A.R.; Dreisinger, D.B. Gold Recovery from Refractory Carbonaceous Ores by Pressure Oxidation, Thiosulfate Leaching and Resin–in–Pulp Adsorption. U.S. Patent 5,785,736, 28 July 1998.
86. Fleming, C.A.; Mcmullen, J.; Thomas, K.G.; Wells, J.A. Recent advances in the development of an alternative to the cyanidation process: Thiosulfate leaching and resin in pulp. *Miner. Metall. Process.* **2003**, *20*, 1–9.
87. Nicol, M.G.; O'Malley, G.P. Recovering gold from thiosulfate leach pulps via ion exchange. *JOM* **2002**, *54*, 44–46. [CrossRef]
88. Nicol, M.J.; O'Malley, G.P. *Cyanide: Social, Industrial and Economic Aspects*; The Minerals, Metals and Materials Society: Warrendale, PA, USA, 2001.
89. Perera, W.N.; Senanayake, G.; Nicol, M.J. Interaction of gold(I) with thiosulfate–sulfite mixed ligand systems. *Inorg. Chim. Acta* **2005**, *358*, 2183–2190. [CrossRef]
90. Jeffrey, M.I.; Hewitt, D.M.; Dai, X.; Brunt, S.D. Ion exchange adsorption and elution for recovering gold thiosulfate from leach solutions. *Hydrometallurgy* **2010**, *100*, 136–143. [CrossRef]
91. Naito, K.; Shieh, M.C.; Okabe, T. Chemical behaviour of low valence sulfur compounds. V. Decomposition and oxidation of tetrathionate in aqueous ammonia solution. *Bull. Chem. Soc. Jpn.* **1970**, *43*, 1372–1376. [CrossRef]
92. Rolia, E.; Chakrabarti, C.L. Kinetics of decomposition of tetrathionate, trithionate and thiosulphate in alkaline media. *Environ. Sci. Technol.* **1982**, *16*, 852–857. [CrossRef] [PubMed]
93. Zhang, H.G.; Dreisinger, D.B. The kinetics for the decomposition of tetrathionate in alkaline solutions. *Hydrometallurgy* **2002**, *66*, 59–65. [CrossRef]
94. Breuer, P.L.; Jeffrey, M.I. The effect of ionic strength and buffer choice on the decomposition of tetrathionate in alkaline solutions. *Hydrometallurgy* **2004**, *72*, 335–338. [CrossRef]
95. Varga, D.; Horvath, A.K. Kinetics and mechanism of the decomposition of tetrathionate ion in alkaline medium. *Inorg. Chem.* **2007**, *46*, 7654–7661. [CrossRef] [PubMed]
96. Ferron, C.J.; Turner, D.W.; Stogran, K. Thiosulfate leaching of gold and silver ores: An old process revisited. In Proceedings of the CIM 100th Annual General Meeting, Montreal, QC, Canada, 3–7 May 1998.

97. Navarro, P.; Vargas, C.; Reveco, V.; Orellana, J. Recovery of gold from ammonia-thiosulfate media with amberlite IRA–410 ionic exchange resin. *Rev. Metal.* **2006**, *42*, 354–366. [CrossRef]
98. Zhang, H.G.; Dreisinger, D.B. Gold Recovery from Thiosulfate Leaching. U.S. Patent 6,632,264, 14 October 2003.
99. Chaparro, M.; Munive, G.; Guerrero, P.; Parga, J.R.; Vazquez, V.; Valenzuela, J.L. Gold adsorption in thiosulfate solution using anionic exchange resin. *J. Multidiscip. Eng. Sci. Technol.* **2015**, *2*, 2159–2163.
100. Clifford, D.; Weber, W.J. The determinants of divalent/monovalent selectivity in anion exchangers. *React. Polym.* **1983**, *1*, 77–89. [CrossRef]
101. Riveros, P.A. Selectivity aspects of the extraction of gold from cyanide solutions with ion-exchange resins. *Hydrometallurgy* **1993**, *33*, 43–58. [CrossRef]
102. Averston, J.; Everest, D.A.; Wells, R.A. Adsorption of gold from cyanide solutions by anionic resins. *J. Chem. Soc.* **1958**, 231–239. [CrossRef]
103. Hardland, C.E. *Ion Exchange: Theory and Practice*; The Royal Society of Chemistry: Cambridge, UK, 1994.
104. Lukey, G.C.; Van Deventer, J.S.J.; Shallcross, D.C. Selective elution of copper and iron cyanide complexes from an ion exchange resins using saline solutions. *Hydrometallurgy* **1999**, *56*, 217–236. [CrossRef]
105. Iguchi, A. The separation of sulfate, sulfite, thiosulfate and sulfite ions with anion-exchange resins. *B. Chem. Soc. Jpn.* **1958**, *31*, 600–605. [CrossRef]
106. Eusebius, L.C.T.; Ghose, A.K.; Mahan, A.; Dey, A.K. Thiosulfate as a complexing agent in the separation of cations by anion-exchange chromatography. *Analyst* **1980**, *105*, 52–59. [CrossRef]
107. Weir, S.I.; Butler, E.C.V.; Haddad, P.R. Ion chromatography with UV detection for the determination of thiosulfate and polythionates in saline waters. *J. Chromatogr. A* **1994**, *671*, 197–203. [CrossRef]

© 2017 by the authors. Licensee MDPI, Basel, Switzerland. This article is an open access article distributed under the terms and conditions of the Creative Commons Attribution (CC BY) license (http://creativecommons.org/licenses/by/4.0/).

Article

Intensification Behavior of Mercury Ions on Gold Cyanide Leaching

Qiang Zhong [1], Yongbin Yang [1], Lijuan Chen [2], Qian Li [1,*], Bin Xu [1] and Tao Jiang [1]

1. School of Minerals Processing and Bioengineering, Central South University, Changsha 410083, China; zhongqiang2008csu@163.com (Q.Z.); ybyangcsu@126.com (Y.Y.); xuandy_16@126.com (B.X.); jiangtao@csu.edu.cn (T.J.)
2. School of Mineral and Environment, Hunan Nonferrous Metals Vocational and Technical College, Zhuzhou 412006, China; csuchenlijuan@163.com
* Correspondence: csuliqian@126.com; Tel.: +86-731-8883-0547

Received: 5 December 2017; Accepted: 17 January 2018; Published: 21 January 2018

Abstract: Cyanidation is the main method used to extract gold from gold raw materials; however, a serious problem with this method is the low leaching rate. In order to improve gold leaching, the intensification behavior of mercury ions on gold cyanide leaching, for two types of materials, sulphide gold concentrate and oxide gold ore, was investigated. The results showed that mercury ions, with only a 10^{-5} M dosage, could significantly intensify leaching and gold recovery. The dissolution behavior of gold plate was also intensified by 10^{-5} M mercury ions. Microstructure analysis showed that mercury ions intensified the cyanidation corrosion of the gold surface, resulting in a loose structure, where a large number of deep ravines and raised particles were evident across the whole gold surface. The loose structure added contact surface between the gold and cyanide, and accelerated gold dissolution. Moreover, mercury ions obstructed the formation of insoluble products, such as AuCN, Au(OHCN), and Au(OH)$_x$, that lead to a passivation membrane on the gold surface, reducing contact between the gold and cyanide. These effects, brought about by mercury ions, change the structure and product of the gold surface during gold cyanidation and promote gold leaching.

Keywords: gold cyanidation; mercury ions; intensification behavior; structure; surface product

---

1. Introduction

Gold is a rare and precious metal. It is not only a special currency for reserve and investment, but also an important material for the jewelry, electronics, and aerospace sectors, among others. Cyanidation has numerous advantages, in contrast with other methods, and is the main method to extract gold from gold materials, such as sulphide gold concentrate and oxide gold ore. However, a serious problem associated with this method is the low leaching rate [1–3]. Through past research, a large number of intensification techniques have been established. Hydrogen peroxide, as an assistant reagent, has been shown to accelerate gold leaching since its appearance in 1987 [4,5].

Gold leaching, in essence, is an electrochemical process that includes the anode dissolution of gold and the cathode reduction of oxygen and other oxidants [6,7]. Hydrogen peroxide assistant leaching is an intensified method for cathode reduction. After cathode reduction is intensified to an extreme level, further intensification no longer contributes to increased gold leaching and the subsequent leaching is determined by anode dissolution. Meanwhile, anodic intensification has received wide attention. Some heavy metals, such as lead, bismuth, silver, and thallium, have been shown to be effective in the intensification of the anodic dissolution of gold [8–11]. Their catalysis on gold leaching has been analyzed theoretically and research has also looked at their intensive electrochemistry.

However, there have been few reports on the practical applications of these technologies and their intensified mechanisms, especially for mercury ions.

Mercury ions can be recycled and reutilized to intensify gold cyanide leaching. After cyanide leaching, gold is separated from the lixivium by active carbon adsorption, and the leaching raffinate, with mercury ions, will circulate back to the previous process and leach gold again. Even if only a few mercury ions exist in the gold concentrate, they will be treated to realize the green discharge of mercury ions. Therefore, there are no pollution problems brought about by mercury intensification in gold cyanide leaching, and it is a suitable intensifier. In our previous research [12–14], the electrochemical kinetics of gold anodic dissolution, intensified by heavy metals, was investigated. Additionally, the co-intensification of heavy metals and hydrogen peroxide to accelerate gold leaching from different types materials was also researched. Moreover, the electrochemical nature of the co-intensification of mercury ions and oxidants was discussed. Some knowledge was obtained based on these works; however, the intensification behavior and mechanism of mercury ions on gold cyanide leaching lack in-depth and systematic understanding. In this work, mercury ions, as the only intensified reagent, were applied to prevent the effect of hydrogen peroxide. A pure gold plate was employed to avoid the disturbance of other elements in the gold material. The leaching behavior of two types of materials, sulphide gold concentrate and oxide gold ore, and the dissolution behavior, structure information, and surface product of the gold plate, were investigated to analyze the intensification of mercury ions on gold cyanide leaching. This research will allow the intensification behavior and mechanism of mercury ions on gold cyanide leaching to be understood more clearly.

2. Materials and Methods

2.1. Materials

Two common gold materials, sulphide gold concentrate and oxide gold ore, were provided by a gold plant in China. As shown in Table 1, sulphide gold concentrate is a high gold-content material (73.21 g/t), while there is only 4.42 g/t gold-content in oxide gold ore. The oxide gold ore has little content of S and Fe; whereas, in the sulphide gold concentrate, the content of S and Fe is much higher at 28.34% and 29.57%, respectively, which may have a negative effect on gold leaching. Moreover, there is 29.57% Fe and 2.19% Cu in the sulphide gold concentrate, which indicates a high content of pyrite and that some copper pyrites exist in the gold concentrate. Meanwhile, a cylindrical gold plate (Φ 7.0 mm × 2.0 mm) with 99.99% purity was employed to analyze the effect of mercury ions on gold dissolution. Analytic reagents, $HgSO_4$ and $NaOH$, were applied as the mercury ion intensifier and solution pH regulator, respectively. NaCN was used as a leaching agent.

Table 1. Chemical composition of gold concentrate (%).

Sample	Au *	Ag *	Cu	Pb	Zn	Fe	S
Sulphide gold concentrate	73.21	176.05	2.19	0.98	1.03	29.57	28.34
Oxide gold ore	4.42	0.36	0.05	0.04	0.05	3.21	0.16

* Unit g/t.

2.2. Methods

2.2.1. Gold Leaching

In the gold leaching test, a 50 g sample was first ground in a Φ 160 mm × 50 mm wet ball mill with a 40% mass fraction in a grinding slurry. After grinding, the slurry was put in a 500 mL breaker and a certain amount of cyanide NaCN, intensifier $HgSO_4$, pH regulator NaOH, and water, was added to the beaker. The concentration of NaCN was 0.3%, the liquid–solid ratio of leaching was 5:1, and the pH of the solution was 12. The concentration of mercury ions was 0, 10^{-6}, 10^{-5}, and 10^{-4} M, as determined by the experimental requirements. The slurry was stirred with a EUROSTAR stepless speed regulation stirrer, with a rotation speed of 400 rpm, and leaching time was recorded. After leaching for a given time, the slurry was filtered and the residue was dried in an electric drying oven (CIE, Changsha, China). The residue and pregnant solution were both analyzed by atomic absorption spectrometry (AA240FS+GTA120, VARIAN, San Francisco, CA, USA) to determine gold content.

2.2.2. Gold Dissolution

The cylindrical gold plate was prepared before cyanide dissolution. Firstly, the gold plate was rubbed and polished by a 0.06-A metallographic sandpaper. Secondly, the gold plate was processed with a high temperature treatment using an alcohol burner. Finally, the gold plate was ultrasonically cleaned for 8 min with successive cleaning solutions of distilled water, 10% diluted nitric acid, distilled water, anhydrous ethanol, and distilled water.

After being prepared, the gold plate was bonded on the stirring rod of a speed regulation stirrer. Then, the stirring rod, along with the gold plate, were placed in a solution with an NaCN concentration of 0.3% and pH of 12. Meanwhile, a certain amount of $HgSO_4$ reagent was added into the solution for Hg intensification cyanidation. There were no mercury ions in the common cyanidation. The stirrer was rotated at a speed of 400 rpm and leaching time was recorded. After leaching for a given time, the gold plate was cleaned using distilled water and naturally air-dried. Afterwards, the gold plate was weighed and its microstructure and surface product were analyzed.

2.2.3. Microstructure and Surface Product Analysis after Dissolution

The microstructure and surface product of the gold plate after common cyanidation and Hg intensification cyanidation were analyzed by scanning electron microscope (SEM), atomic force microscope (AFM), X-ray photoelectron spectrometer (XPS), and fourier transform infrared spectoscopy (FT-IR) (NEXUS 670, NICOLET, Wisconsin Rapids, WI, USA). SEM analysis was conducted using a JSM-6360LV scanning electron microscope (Quanta 250 FEG, FEI, Hillsboro, OR, USA). AFM analysis was conducted using a di NanoMan Vs AFM atomic force microscope (Multimode VIII, VEECO, Plainview, NY, USA). XPS analysis was conducted using a K_a 1063 X-ray photoelectron spectrometer (AXIS ULTRA DLD, Shimadzus, Kyoto, Japan), whose vacuum degree, X-ray source, and energy were 1×10^{-7} Pa, Al K_a, and 50 eV, respectively. Infrared spectrum analysis was performed using a Nexsus 670 IR spectrometer, with a resolution of 0.09 cm^{-1} and scan time of 60 times/s.

3. Results

3.1. The Effect of Mercury Ions on Gold Leaching

3.1.1. Sulphide Gold Concentrate

As a common gold concentrate in China, sulphide gold concentrate was used to investigate the intensification behavior of mercury ions on gold leaching. As shown in Figure 1, mercury ions have an obvious positive effect on sulphide gold concentrate leaching, especially for 10^{-5} and 10^{-4} M mercury ions. Mercury ions not only increase the gold leaching rate, but also improve final gold recovery. After adding 10^{-6} M Hg, gold recovery increased from 65.7% to 70.1% when the leaching time was two hours. Continuing to extend leaching time, gold recovery with Hg intensification was also higher than with common cyanidation. With 10^{-5} M Hg, gold recovery showed an obvious improvement compared to both common cyanidation and 10^{-6} M Hg intensification. With 10^{-5} M Hg, gold recovery was up to 77.6% after only two hours of leaching, while the same recovery with common cyanidation took nearly eight hours. Meanwhile, recovery further improved with extended leaching time and recovery after 24 hours leaching reached a high of 85.8%. Increasing the content of mercury ions to 10^{-4} M improved gold recovery, but this recovery was not considerably different to 10^{-5} M Hg. Therefore, mercury ions were an effective intensifier for the cyanide leaching of sulphide gold concentrate. Only with 10^{-5} M Hg can the gold cyanide leaching be adequately intensified.

In addition, both common cyanidation and Hg intensification cyanidation took nearly twelve hours to complete gold leaching.

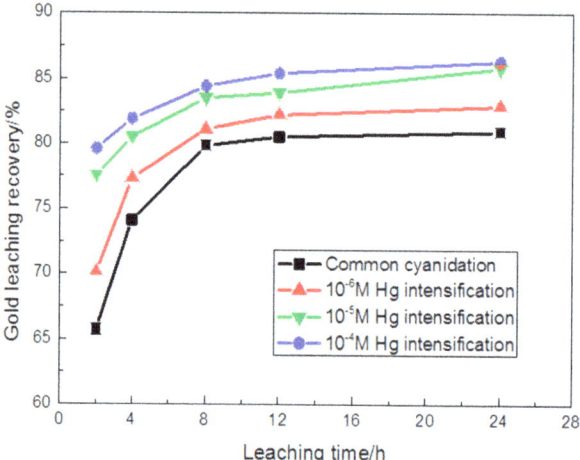

Figure 1. The effect of mercury ions with different content on gold leaching (in sulphide gold concentrate).

3.1.2. Oxide Gold Ore

In the same way, the intensification behavior of mercury on the gold leaching of oxide gold ore, was investigated. As shown in Figure 2, the leaching behavior of oxide gold ore was similar to that of sulphide gold concentrate. With 10^{-6} M Hg, gold recovery was 73.5% after two hours, compared to recovery of 68.4% with common cyanidation. With 10^{-6} M Hg, recovery reached 80.4% after four hours of leaching. Increasing the content of mercury ions to 10^{-5} and 10^{-4} M improved gold recovery and the leaching patterns were almost the same. Gold recovery with 10^{-5} M Hg intensification was as high as 81.3%, 84.2%, and 87.1% after leaching for two, four and twelve hours, respectively. Therefore, the gold cyanide leaching of oxide gold ore was intensified after adding 10^{-5} M Hg. Moreover, as with sulphide gold concentrate, oxide gold ore also needed twelve hours to complete its gold leaching.

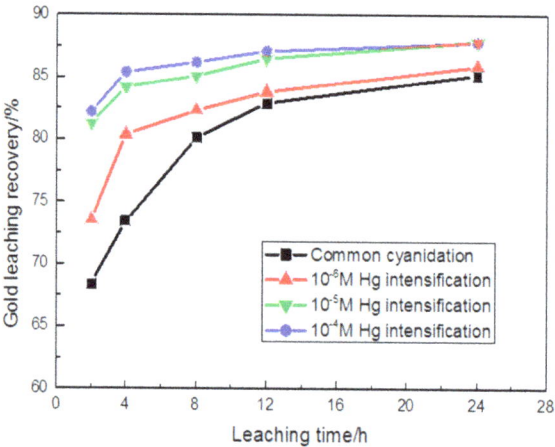

Figure 2. The effect of mercury ions with different content on gold leaching (in oxide gold ore).

3.2. The Effect of Mercury Ions on Gold Dissolution

In order to avoid other factors and to better research the effect of mercury ions on gold leaching, a pure gold plate was used to analyze the intensification behavior of mercury ions on gold dissolution. The content of the mercury ions was 10^{-5} M and the effect at different time points on gold plate dissolution, was measured. As shown in Figure 3, the quality loss of the gold plate during Hg intensification cyanidation was far greater than that of common cyanidation. After three hours of dissolution, the quality loss of the gold plate with Hg intensification was 0.489 mg, while the quality loss with the common dissolution was only 0.113 mg. Even with 12 hours of common dissolution, the quality loss of the gold plate was 0.421 mg, which was still lower than after three hours of Hg intensification cyanidation. Moreover, the quality loss of the gold plate after 12 hours of Hg intensification dissolution was 1.487 mg. This shows that gold dissolution can be intensified by mercury ions. Similarly, mercury ions can promote the removal of gold from gold materials and promote gold dissolution in a cyanide solution.

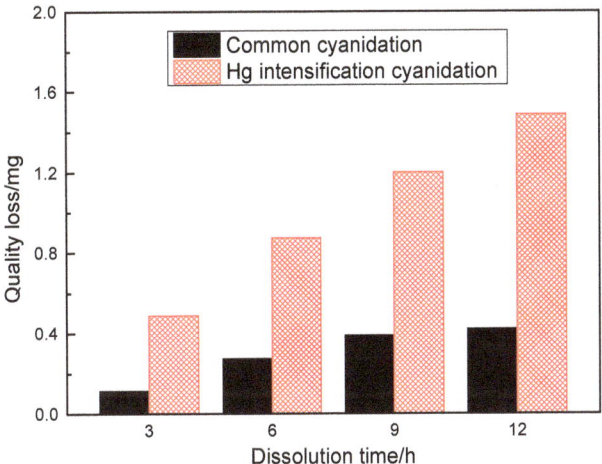

Figure 3. Gold dissolution at different time points.

3.3. Microstructure after Dissolution

3.3.1. Structure Information Determined by SEM

The original gold plates were analyzed using SEM, as well as the same gold plates after six hours of cyanide dissolution (common or Hg intensification). As shown in Figure 4A1,A2, the microstructures of the two gold plates were almost the same, which means that the surface structures of the gold plates showed little change during common cyanide dissolution. However, the surface structure of the gold plates differed substantially after Hg intensification dissolution, as shown in Figure 4B1,B2. With the original gold plate, there were some shallow ravines on the surface, but it was an integral whole with a homogeneous structure. After Hg intensification dissolution, many deep ravines and pits appeared, making the surface rough with a loss of structure. Gold plates can be severely corroded and gold is removed from plates during Hg intensification dissolution. Meanwhile, the rough surface and loose structure improves the contact area between the gold and cyanide, further accelerating gold dissolution. Therefore, mercury ions have an important effect on gold plate structure. Hg intensification cyanidation corrodes the gold surface, destroys its structure, and ultimately accelerates gold dissolution.

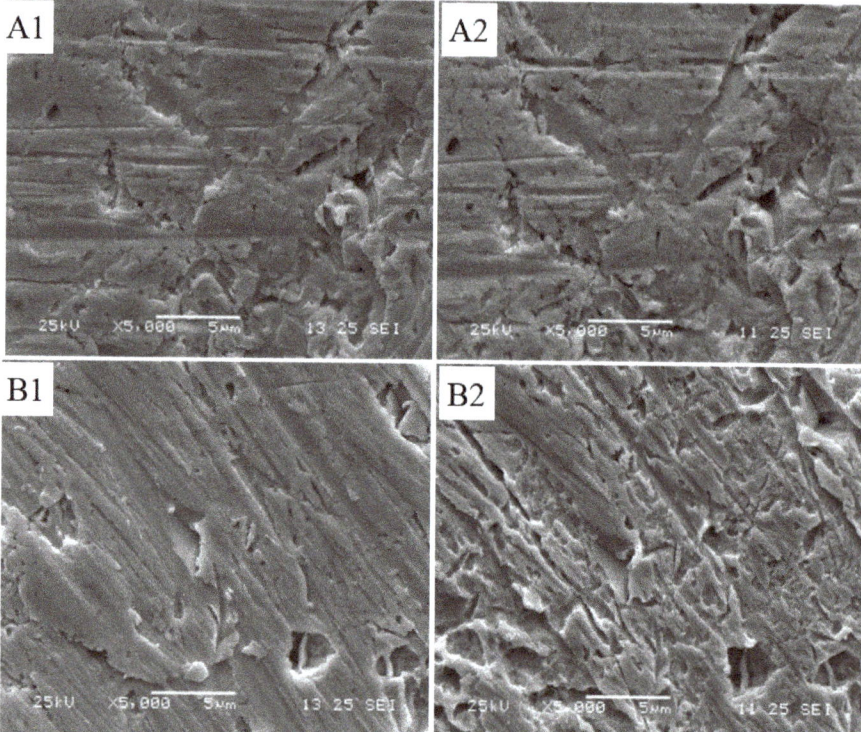

Figure 4. The microstructure of different gold surfaces, as shown by SEM; (**A1**, **B1**) The original gold plates; (**A2**) The gold plate after common cyanidation; (**B2**) The gold plate after Hg intensification cyanidation.

Meanwhile, the original gold plate was put into the same Hg cyanide solution and dissolved in a stationary state. The gold plate, after different time points of dissolution, was analyzed using SEM, and the structure information from this analysis is shown in Figure 5. The surface of the original gold plate was smooth and homogeneous. After one day of dissolution, there were many shallow pits on the surface of the gold plate, and while the hole displayed some changes, these were not obvious when compared to the original gold plate. After two days of dissolution, the whole surface was rough and blurry, which may have been due to cyanide corrosion on the gold. Meanwhile, the hole displayed an obvious change, where the area of the hole was larger than that of Figure 5A,B, and some new holes had formed, as shown in Figure 5C. After three days of dissolution, the original hole displayed no obvious change, but the new holes were larger and deeper. Continuing to extend the dissolution time, the structure of the gold plate showed no obvious change. The results show that it should take three days to complete gold dissolution, which is uneconomical and unrealistic. Therefore, the intensification of mercury ions on gold dissolution is limited and inefficient when gold is dissolved in a stationary state. The intensification must combine with stirring.

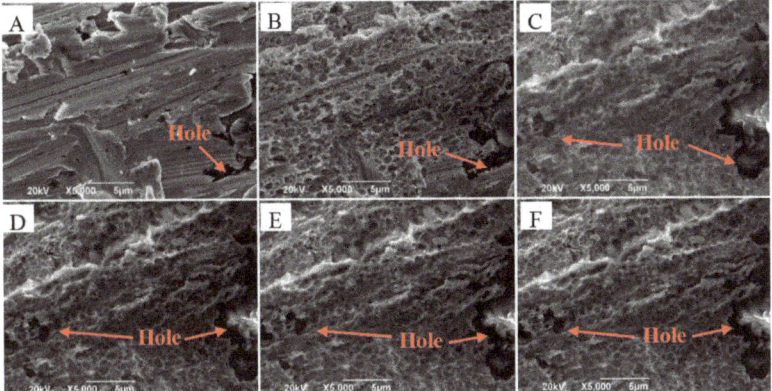

Figure 5. The microstructure of the gold surface, as shown by SEM; (**A**) The original gold plate; (**B–F**) dissolution after one day, two days, three days, four days, and five days, respectively.

3.3.2. Structure Information Determined by Atomic Force Microscope (AFM)

To obtain more accurate structural information about the gold plates, the same gold plates, after SEM analysis, were analyzed using AFM. As shown in Figure 6, the structure of the three gold plates displayed different characteristics. The surface of the original gold plate was smooth and homogeneous, while its surface was rough and rugged after cyanidation, implying that the gold plate was corroded by the cyanide. Cyanide corrosion changed the surface structure of the gold plate, especially after Hg intensification cyanidation, where a large number of raised particles were evident across the whole surface. This structure change can be visually observed in the three-dimensional structure drawing in Figure 6A2,B2,C2. In contrast to the surface structure of the gold after common cyanidation, there were many overlapping, raised particles on the gold surface after Hg intensification cyanidation, making the structure rougher and more rugged than that of common cyanidation. This means that Hg intensification cyanidation makes gold plates suffer more serious cyanide corrosion than common cyanidation. In the same way, mercury ions can promote the removal of gold from gold materials and intensify gold cyanide dissolution.

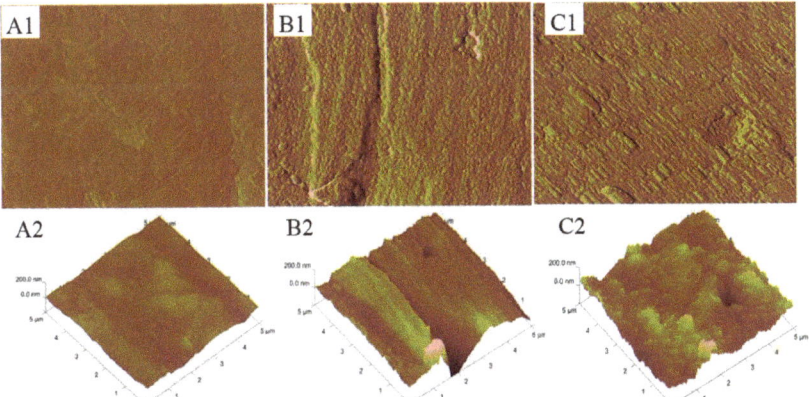

Figure 6. The microstructure of different gold surfaces, as shown by AFM; (**A1,A2**) The original gold plate; (**B1,B2**) The gold plate after common cyanidation; (**C1,C2**) The gold plate after Hg intensification cyanidation.

3.4. Surface Product after Dissolution

Gold cyanide dissolution can be modeled by the following simplified equation [13,15,16]:

$$Au + CN^- = AuCN + e \quad (1)$$

$$AuCN + CN^- = Au(CN)_2^- \quad (2)$$

In the dissolution, the gold is the first cyanide resolved to the intermediate product, AuCN, which is insoluble in water, then the AuCN further reacts with CN^- and generates the final product of $Au(CN)_2^-$, which is soluble in water. Meanwhile, the following equation occurs in the process of gold dissolution [17–20]:

$$AuCN + OH^- = Au(OHCN) + e \quad (3)$$

$$AuCN + X \cdot OH^- = Au(OH)_x + CN^- + (X-1) \cdot e \quad (4)$$

AuCN, Au(OHCN), and $Au(OH)_x$, all insoluble in water, are the main intermediate products in cyanide dissolution. In order to reveal the products on the gold plate surface, such as AuCN, Au(OHCN), and $Au(OH)_x$, and their quantitative relationship, the gold plates were analyzed after cyanide dissolution (common and Hg intensification).

3.4.1. Surface Product Information Determined by X-ray Photoelectron Spectrometer (XPS)

After cyanide dissolution (common and Hg intensification), the gold plates were analyzed by XPS to obtain molecular level product information regarding their surface after cyanidation. The results of the XPS spectra (Au4f, C1s, N1s, O1s, and Hg4f) are shown in Figure 7. Both of the XPS spectra were similar, except that there was a weak Hg4f peak in the spectra of the Hg intensification, which was caused by the addition of mercury ions. The similar spectra meant that the surfaces of the two gold plates were composed of similar chemical elements. That the Au4f peak was larger than the others peak indicates that Au is the main element of the gold plate surface.

To reveal the surface product differences between the two gold plates, the spectra of Au4f, C1s, N1s, and O1s, were analyzed separately, as shown in Figure 8. Both of the Au4f spectra showed two peaks, while the C1s, N1s, and O1s spectra each exhibited only one peak. For the two Au spectra, the peaks with lower binding energy, at 83.58 and 83.78 eV, likely originated from Au, and the peaks with higher binding energy, at 87.28 and 87.48 eV, may be attributed to Au^- in the form of AuCN [21,22]. As shown by the N1s spectra, the binding energy of common cyanidation and Hg intensification cyanidation was 399.38 and 399.48 eV, respectively, which may have been derived from CN^- in the form of AuCN [23,24]. As shown by the O1s spectra, the binding energy of common cyanidation and Hg intensification cyanidation was 531.78 and 531.58 eV, respectively, which likely originated from OH^- in the form of $Au(OH)_x$ and Au(OHCN) [25]. Meanwhile, as shown by the C1s spectra, the binding energy of common cyanidation and Hg intensification cyanidation was 284.88 and 284.78 eV, respectively, which may be attributed to graphite carbon [26]. According to the comparison of peak areas, it is known that the quantities of CN^- and OH^- on the gold plate surface, are distinctly greater after common cyanidation than Hg intensification cyanidation. This means that a higher content of AuCN, $Au(OH)_x$, and Au(OHCN) existed on the surface of the gold plate after common cyanidation compared to after Hg intensification cyanidation.

Moreover, the C1s resolved spectra of the two gold plates were analyzed, as shown in Figure 9. The peaks with lower binding energy, at 284.78 and 284.88 eV, may have originated from graphite carbon that was an insignificant component during the test [26]. The peaks with higher binding energy, at 287.98 and 288.19 eV, are likely attributed to CN^- in the form of AuCN [23,24,26]. The results for C1s-resolved spectra support the findings of the N1s XPS analysis.

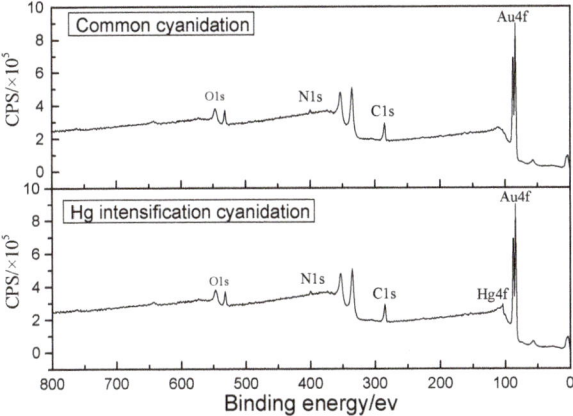

Figure 7. XPS spectra of the gold plates after cyanidation (common and Hg intensification).

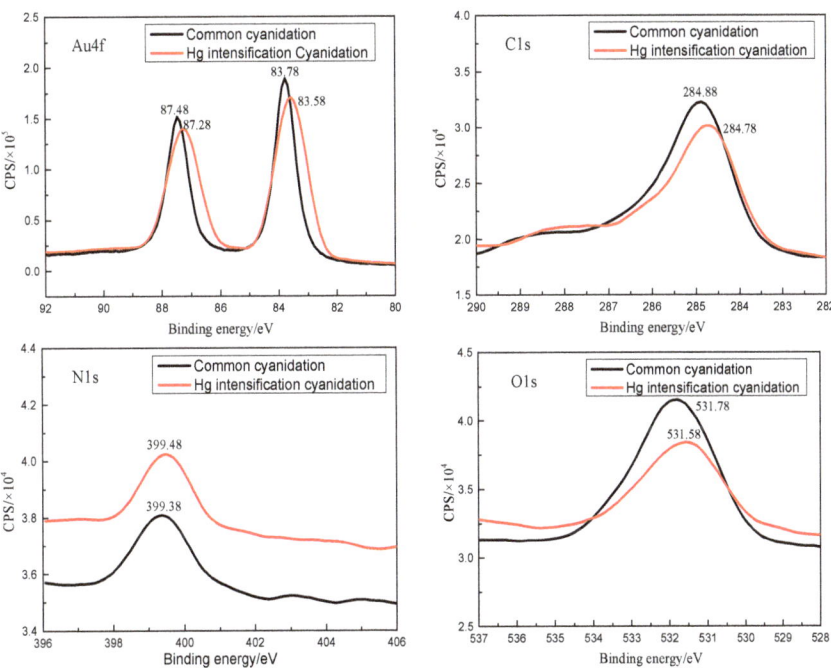

Figure 8. Au4f, C1s, N1s, and O1s, XPS spectra of the gold plates after cyanidation (common and Hg intensification).

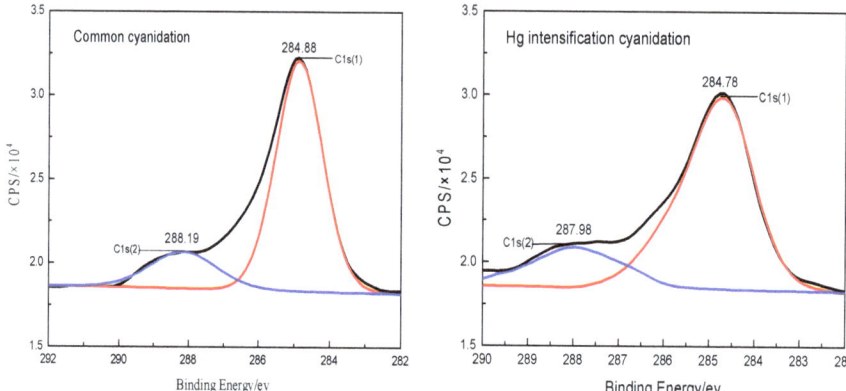

Figure 9. C1s XPS-resolved spectra of the gold plates after cyanidation (common and Hg intensification).

3.4.2. Surface Product Information Determined by Fourier Transform Infrared Spectroscopy (FT-IR)

After XPS analysis, the same gold plates were analyzed using FI-IR. The results of this analysis are shown in Figure 10 and Table 2. Comparing the two spectra, their curves showed some similarities. The peaks at 1240 and 1650 cm^{-1}, that were evident on both of the spectra, were alcohol compounds and nitro compounds, respectively [27–29]. These may have been the residues of the nitric acid and ethanol used in the cleaning process of the gold plate. The peaks at 2850 and 2920 cm^{-1}, that were also evident on both of the spectra, were hydrocarbon compounds [27,28,30]. The hydrocarbon compounds were a mixed impurity introduced during the cyanide dissolution because the original composition was Au, NaOH, NaCN, H$_2$O, and HgSO$_4$. The peaks at 3350 and 3460 cm^{-1} were alkaline compounds [27,28,30], caused by sodium hydroxide. Moreover, there was a peak at 2100 cm^{-1} on the spectrum of common cyanidation. This was the cyanide compound (C≡N) [31,32], meaning that considerable amounts of cyanide compound were evident on the gold plate surface after common cyanidation. Based on Equations (1)–(4), the Au(CN)$_2^-$ product is soluble, which indicates that it will be removed from the gold plate surface and dissolve into the cyanide solution. The insoluble products, AuCN and Au(OHCN), cannot dissolve into the cyanide solution so deposit on the gold plate surface [33]. Therefore, the considerable cyanide compounds that existed on the surface of the gold plates after common cyanidation were AuCN and Au(OHCN). This peak (C≡N) did not exist on the spectrum of Hg intensification dissolution, indicating that a large number of cyanide compounds were not evident on the surface. This means mercury ions can promote the conversion of AuCN to Au(CN)$_2^-$ and obstruct the generation of Au(OHCN), preventing the deposition of AuCN and Au(OHCN) on the gold surface.

Comprehensive analysis of the XPS and FT-IR results showed that the insoluble intermediate products, AuCN, Au(OHCN), and Au(OH)$_x$, were constantly produced and deposited on the gold surface, forming a passivation membrane, which obstructs the reaction of gold and cyanide during common cyanidation. Mercury ions promoted the conversion of AuCN to Au(CN)$_2^-$ and obstructed the generation of Au(OHCN) and Au(OH)$_x$, which prevented their deposition on the gold surface, maintained good contact between the gold and cyanide, and, finally, ensured the gold was adequately leached by the cyanide.

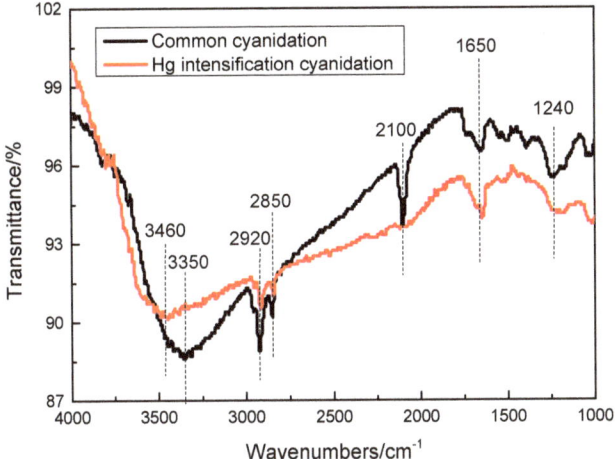

Figure 10. Infrared spectra of the different gold plates.

Table 2. FT-IR (Fourier Transform Infrared spectroscopy) spectra of the gold plates under different conditions.

Position (cm^{-1})	Assignment	Group
1240	O–H	Alcohol compound
1650	NO$_2$	Nitro compound
2100	C≡N	Cyanide compound
2850, 2920	C–H	Hydrocarbon compound
3460, 3550	O–H	Alkaline compound

4. Conclusions

The leaching behavior of two types of materials, sulphide gold concentrate and oxide gold ore, and the dissolution behavior, structure information, and surface product of gold plates, were investigated to analyze the intensification of mercury ions on gold cyanide leaching. A summary of the results obtained in this study is as follows:

(1) Mercury ions intensified the leaching of sulphide gold concentrate and oxide gold ore, and gold recovery was significantly improved. Meanwhile, mercury ions could be recycled and reutilized to intensify gold cyanide leaching, with no pollution problems brought about by mercury intensification. After adding 10^{-5} M Hg, gold recovery was about 80% after only two hours of leaching and reached close to 90% after 12 hours leaching. Similarly, the pure gold plate was also intensified by mercury ions.

(2) Mercury ions had an obvious effect on the surface structure of the gold plate during cyanide dissolution. Mercury ions intensified the cyanidation corrosion on the gold surface and destroyed its structure, resulting in a large number of deep ravines and raised particles covering the whole surface. This loose structure added to the surface contact area between the gold and cyanide, accelerating the gold dissolution.

(3) With common cyanidation, the insoluble intermediate products of AuCN, Au(OHCN), and Au(OH)$_x$ were constantly deposited onto the gold surface, forming a passivation membrane that obstructed the reaction of gold and cyanide. Mercury ions promoted the conversion of AuCN to Au(CN)$_2^-$ and obstructed the generation of Au(OHCN) and Au(OH)$_x$, which prevented their deposition on the gold surface, promoted good contact between the gold and cyanide, and, finally, ensured the gold was adequately leached by the cyanide.

Acknowledgments: This work was supported by the National Natural Science foundation of China (No. 51574284 and No. 51504293) and the Postdoctoral Science Foundation of Central South University.

Author Contributions: Qiang Zhong, Yongbin Yang, and Lijuan Chen conceived and designed the experiments; Qiang Zhong performed the experiments; Yongbin Yang and Lijuan Chen analyzed the data; Qian Li and Bin Xu contributed reagents, materials, and analysis tools; Qiang Zhong and Qian Li wrote the paper; and Tao Jiang reviewed it before submission.

Conflicts of Interest: The authors declare no conflict of interest.

References

1. Bas, A.D.; Ghali, E.; Choi, Y. A review on electrochemical dissolution and passivation of gold during cyanidation in presence of sulphides and oxides. *Hydrometallurgy* **2017**, *172*, 30–44. [CrossRef]
2. Bas, A.D.; Zhang, W.; Ghali, E.; Choi, Y. A study of the electrochemical dissolution and passivation phenomenon of roasted gold ore in cyanide solutions. *Hydrometallurgy* **2015**, *158*, 1–9. [CrossRef]
3. Acar, S. Process development metallurgical studies for gold cyanidation process. *Miner. Metall. Proc.* **2016**, *33*, 161–171. [CrossRef]
4. Guzman, L.; Segarra, M.; Chimenos, J.M.; Fernandez, M.A.; Espiell, F. Gold cyanidation using hydrogen peroxide. *Hydrometallurgy* **1999**, *52*, 21–35. [CrossRef]
5. Nunan, T.O.; Viana, I.L.; Peixoto, G.C.; Ernesto, H.; Verster, D.M.; Pereira, J.H.; Bonfatti, J.M.; Teixeira, L.A.C. Improvements in gold ore cyanidation by pre-oxidation with hydrogen peroxide. *Miner. Eng.* **2017**, *108*, 67–70. [CrossRef]
6. Lin, H.K.; Chen, X. Electrochemical study of gold dissolution in cyanide solution. *Miner. Metall. Proc.* **2001**, *18*, 147–153.
7. Yang, L.; Jia, F.; Song, S. Recovery of $[Au(CN)_2]^-$ from gold cyanidation with graphene oxide as adsorbent. *Sep. Purif. Technol.* **2017**, *186*, 63–69. [CrossRef]
8. Guzman, L.; Chimenos, J.M.; Fernandez, M.A.; Segarra, M.; Espiell, F. Gold cyanidation with potassium persulfate in the presence of a thallium (I) salt. *Hydrometallurgy* **2000**, *54*, 185–193. [CrossRef]
9. Tshilombo, A.F.; Sandenbergh, R.F. Electrochemical study of the effect of lead and sulphide ions on the dissolution rate of gold in alkaline cyanide solutions. *Hydrometallurgy* **2001**, *60*, 55–67. [CrossRef]
10. Deschenes, G.; Lastra, R.; Brown, J.R.; Jin, S.; May, O.; Ghali, E. Effect of lead nitrate on cyanidation of gold ores: Progress on the study of the mechanisms. *Miner. Eng.* **2000**, *13*, 1263–1279. [CrossRef]
11. Jeffrey, M.I.; Ritchie, I.M. The leaching of gold in cyanide solutions in the presence of impurities II. The effect of silver. *J. Electrochem. Soc.* **2000**, *147*, 3272–3276. [CrossRef]
12. Yang, Y.B.; Li, Q.; Li, G.H.; Guo, Y.F.; Huang, Z.C.; Jiang, T. An electrochemical investigation on intensification of gold cyanidation by heavy metal ions. In *EPD Congress 2005*; Schlesinger, M.E., Ed.; TMS (The Minerals, Metals & Materials Society): Pittsburgh, PA, USA, 2005; pp. 977–984.
13. Li, Q.; Jiang, T.; Yang, Y.B.; Li, G.H.; Guo, Y.F.; Qiu, G.Z. Co-intensification of cyanide leaching gold by mercury ions and oxidant. *Trans. Nonferr. Met. Soc.* **2010**, *20*, 1521–1526. [CrossRef]
14. Yang, Y.B.; Li, Q.; Jiang, T.; Guo, Y.F.; Li, G.H.; Xu, B. Co-intensification of gold leaching with heavy metals and hydrogen peroxide. *Trans. Nonferr. Met. Soc.* **2010**, *20*, 903–909. [CrossRef]
15. Arthur, D.M.M. A study of gold reduction and oxidation in aqueous solutions. *J. Electrochem. Soc.* **1972**, *119*, 672–677. [CrossRef]
16. Cathro, K.J.; Koch, D.F.A. The anodic dissolution of gold in cyanide solutions. *J. Electrochem. Soc.* **1964**, *111*, 1416–1420. [CrossRef]
17. Li, J. Electrochemical study of silver dissolution in cyanide solutions. *J. Electrochem. Soc.* **1993**, *140*, 1921–1927. [CrossRef]
18. Pan, T.P.; Wan, C.C. Anodic behaviour of gold in cyanide solution. *J. Appl. Electrochem.* **1979**, *9*, 653–655. [CrossRef]
19. Eisenmann, E.T. Kinetics of the electrochemical reduction of dicyanoaurate. *J. Electrochem. Soc.* **1978**, *125*, 717–723. [CrossRef]
20. Sandenbergh, R.F.; Miller, J.D. Catalysis of the leaching of gold in cyanide solutions by lead, bismuth and thallium. *Miner. Eng.* **2015**, *14*, 1379–1386. [CrossRef]

21. Moulder, J.F.; Chastain, J.; King, R.C. Handbook of x-ray photoelectron spectroscopy: A reference book of standard spectra for identification and interpretation of XPS data. *Chem. Phys. Lett.* **1995**, *220*, 7–10.
22. Bas, A.D.; Safizadeh, F.; Zhang, W.; Ghali, E.; Choi, Y. Active and passive behaviors of gold in cyanide solutions. *Trans. Nonferr. Met. Soc.* **2015**, *25*, 3442–3453. [CrossRef]
23. Ma, L.; Fan, H.; Wang, J.; Zhao, Y.; Tian, H.; Dong, G. Water-assisted ions in situ intercalation for porous polymeric graphitic carbon nitride nanosheets with superior photocatalytic hydrogen evolution performance. *Appl. Catal. B-Environ.* **2016**, *190*, 93–102. [CrossRef]
24. Obrosov, A.; Gulyaev, R.; Ratzke, M.; Volinsky, A.A.; Bolz, S.; Naveed, M.; Weiß, S. XPS and AFM investigations of Ti-Al-N coatings fabricated using DC magnetron sputtering at various nitrogen flow rates and deposition temperatures. *Metals* **2017**, *7*, 52. [CrossRef]
25. Štrbac, S.; Smiljanić, M.; Rakočević, Z. Spontaneously deposited Rh on Au(111) observed by AFM and XPS: Electrocatalysis of hydrogen evolution. *J. Electrochem. Soc.* **2016**, *163*, 3027–3033. [CrossRef]
26. Goff, A.L.; Artero, V.; Metayé, R.; Moggia, F.; Jousselme, B.; Razavet, M.; Tran, P.D.; Palacin, S.; Fontecave, M. Immobilization of FeFe hydrogenase mimics onto carbon and gold electrodes by controlled aryldiazonium salt reduction: An electrochemical, XPS and ATR-IR study. *Int. J. Hydrog. Energy* **2010**, *35*, 10790–10796. [CrossRef]
27. Geng, W.; Nakajima, T.; Takanashi, H.; Ohki, A. Analysis of carboxyl group in coal and coal aromaticity by Fourier transform infrared (FT-IR) spectrometry. *Fuel* **2009**, *88*, 139–144. [CrossRef]
28. Barroso-Bogeat, A.; Alexandre-Franco, M.; Fernández-González, C.; Gómez-Serrano, V. FT-IR analysis of pyrone and chromene structures in activated carbon. *Energy Fuels* **2014**, *28*, 4096–4103. [CrossRef]
29. Ibrahim, K.A. Synthesis and characterization of some new aromatic diamine monomers from oxidative coupling of anilines and substituted aniline with 4-amino-N,N-dimethyl aniline. *Arab. J. Chem.* **2014**, *7*, 1017–1023. [CrossRef]
30. Zhong, Q.; Yang, Y.B.; Jiang, T.; Li, Q.; Xu, B. Xylene activation of coal tar pitch binding characteristics for production of metallurgical quality briquettes from coke breeze. *Fuel Process. Technol.* **2016**, *148*, 12–18. [CrossRef]
31. Shriver, D.F.; Shriver, S.A.; Anderson, S.E. Ligand field strength of the nitrogen end of cyanide and structures of cubic cyanide polymers. *Inorg. Chem.* **1965**, *4*, 725–730. [CrossRef]
32. Tseng, T.F.; Yang, Y.L.; Lin, Y.J.; Lou, S.L. Effects of electric potential treatment of a chromium hexacyanoferrate modified biosensor based on PQQ-dependent glucose dehydrogenase. *Sensors* **2010**, *10*, 6347–6360. [CrossRef] [PubMed]
33. Kirk, D.W.; Foulkes, F.R. Anodic dissolution of gold in aqueous alkaline cyanide solutions at low overpotentials. *J. Electrochem. Soc.* **1980**, *127*, 1993–1997. [CrossRef]

© 2018 by the authors. Licensee MDPI, Basel, Switzerland. This article is an open access article distributed under the terms and conditions of the Creative Commons Attribution (CC BY) license (http://creativecommons.org/licenses/by/4.0/).

Article

Selective Extraction of Rare Earth Elements from Phosphoric Acid by Ion Exchange Resins

Xavier Hérès [1],*, Vincent Blet [1], Patricia Di Natale [1], Abla Ouaattou [2], Hamid Mazouz [2], Driss Dhiba [2] and Frederic Cuer [1]

[1] French Nuclear and Alternative Energies Commission (CEA), Nuclear Energy Division—CEA Marcoule, Research Department of Mining and Fuel Recycling ProCesses (DMRC), BP 17171, F-30207 Bagnols sur Ceze, France; Vincent.blet@cea.fr (V.B.); patricia.di-natale@cea.fr (P.D.N.); frederic.cuer@cea.fr (F.C.)
[2] Research & Development Direction, OCP SA., BP 118, Jorf Lasfar El Jadida, El Jadida 24000, Morocco; Abla.OUAATTOU@ocpgroup.ma (A.O.); H.MAZOUZ@ocpgroup.ma (H.M.); D.DHIBA@ocpgroup.ma (D.D.)
* Correspondence: xavier.heres@cea.fr; Tel.: +33-466-797-700

Received: 28 June 2018; Accepted: 25 August 2018; Published: 30 August 2018

Abstract: Rare earth elements (REE) are present at low concentrations (hundreds of ppm) in phosphoric acid solutions produced by the leaching of phosphate ores by sulfuric acid. The strongly acidic and complexing nature of this medium, as well as the presence of metallic impurities (including iron and uranium), require the development of a particularly cost effective process for the selective recovery of REE. Compared to the classical but costly solvent extraction, liquid-solid extraction using commercial chelating ion exchange resins could be an interesting alternative. Among the different resins tested in this paper (Tulsion CH-93, Purolite S940, Amberlite IRC-747, Lewatit TP-260, Lewatit VP OC 1026, Monophos, Diphonix,) the aminophosphonic IRC-747, and aminomethylphosphonic TP-260 are the most promising. Both of them present similar performances in terms of maximum sorption capacity estimated to be 1.8 meq/g dry resin and in adsorption kinetics, which appears to be best explained by a moving boundary model controlled by particle diffusion.

Keywords: phosphoric acid; rare-earth elements; separation; solid-liquid extraction; ion exchange resin

---

1. Introduction

As part of the framework agreement of R&D collaboration between OCP SA and CEA, a study was conducted to evaluate the interest of the recovery of rare earth elements (REE) as by-product from the production of phosphoric acid from phosphate rocks. The leaching of the phosphorite ore with concentrated sulfuric acid (98 vol.%) leads to phosphoric acid solutions with a concentration ranging from 4 to 5 mol/L, whose temperature can range from 60 °C to 70 °C, after phosphogypsum filtration. These filtered solutions contain rare earth elements (REE) (hundreds of ppm) that can be recovered if the economic balance is positive. Among separation techniques, the liquid-liquid extraction has been historically implemented because of the large flow rates involved. However, solid-phase extraction based on chelating ion exchange resins may be of interest for the following reasons:

- There is no risk of contamination of phosphoric acid with organic solvents, so no need for expensive post-acid treatment downstream of extraction [1].
- Solid supports are generally heat resistant [2,3]. This characteristic is very important because of relatively high operating temperatures of phosphoric acid solutions (40–50 °C after storage).

For low feed concentrations, cost of implementation (capital and operating costs) may be lower than in the case of solvent extraction that requires hundreds of mixer-settlers...

Some drawbacks could nevertheless be outlined [4]:

- Resins may be sensitive to chemical degradation in concentrated phosphoric acid.
- Phenomena of swelling may lead to mechanical stresses during sorption-desorption.
- There is a risk of resin poisoning with some cationic impurities in large quantities such as iron, leading to a decrease in sorption capacity and/or selectivity.

Despite these possible limitations, a number of resins have been commercialized, making REE sorption a potential attractive alternative. As an illustration, the cost of an ion exchange (IX) process has recently been compared to that of a solvent extraction (SX) process in the case of uranium recovery from phosphoric acid [2]. Operating costs appear to be much lower for IX technology than those induced by SX ($18 instead of $32 for the production of one pound of U_3O_8). However, this IX process has not been implemented at a commercial scale yet, but rather at a pilot scale for the extraction of uranium in phosphoric acid with sulfonic resins [5] or for the recovery of REE from leachates of uranium ores with sulfonic and phosphonic resins [6].

From an engineering point of view, the resin has to exhibit specific performances for an industrially viable IX extraction of REE from wet phosphoric acid:

- A high selectivity between REE and cationic impurities such as iron (Fe^{3+}), uranium (U(IV), U(VI)) or thorium (Th(IV)), taking into account that Fe^{3+} is approximately 30 to 40 times more concentrated than REE in phosphoric acid solutions. Due to a similarity in extraction chemistry, Fe^{3+} and Th are often adsorbed onto chelating resins in preference to REE [7]. It should be noted that a first precipitation of Fe^{3+} is not an option since it could generate high losses in REE by co-precipitation [6].
- A good capability to bind the REE in highly concentrated H_3PO_4 (4 to 5 mol/L). A total ion exchange capacity greater than 2 meq/g of resin is generally sought (1 eq = 1 mole/valence). The higher the capacity the smaller the process and thus the investment cost.
- The kinetics of sorption and elution should be fast enough to limit the size of the columns.
- A mechanical, physical, and chemical stability over several cycles of extraction/elution will reduce the operating cost.

It is well known [8] that these performances are strongly related to:

- The nature of the matrix (copolymer), the structure and the degree of crosslinking, the nature and the number of fixed ionic (functional) groups.
- The presence of competitive cationic impurities such as U, Th, Fe^{3+}, Al^{3+}, Ca^{2+}, Mg^{2+}, ... In that study it should be emphasized that, no U or Th is present in the OCP's genuine acid solutions due to a preliminary proprietary treatment.
- The nature and pH of the bulk acidic medium.

Based on these remarks, commercially available ion exchange resins have been selected for their potential capability to extract REE from high concentrated phosphoric acid. Some resins have already been tested in close conditions [9] but they exhibit a low yield of extraction (<20%) when the phosphoric acid concentration is 5 mol/L. Moreover, most of the literature concerns other acidic media (hydrochloric, sulfuric, nitric ...) and cannot be directly applied to such an acidic medium containing strong complexing phosphates. In that context, this paper aims at comparing promising commercial resins in view of the industrially viable valorization of REE contained in the highly concentrated phosphoric acid produced by OCP.

2. Choice of the Resin

Numerous functional groups such as dialkylphosphoric, phosphonic, phosphoric, iminodiacetic acids are known to be good REE chelatants [7,10]. Resins with similar functional groups are thus particularly interesting. More specifically, phosphorus-containing resins are widely used for preconcentration of transition metals, lanthanides, uranium, and thorium [11]. These cation exchangers contain either phosphate-OPO (OH) or phosphonic-PO (OH) or phosphinic-PO (OH) H function. The stability of the aminophosphonic group was confirmed with Tulsion CH-93 resin: after 7 cycles of gadolinium (GdIII) extraction in phosphoric acid followed by quantitative elution with ammonium oxalate, Gd^{3+} extracted at the same yield of 70 ± 1% [12]. Another advantage is the low sensitivity of phosphorus resins like Tulsion-CH-96 (phosphinic acid) or T-PAR (phosphoric acid) to the temperature [13]. Increasing the phosphoric acid temperature from 30 °C to 70 °C hardly modifies the REE extraction. Since this phenomenon has been observed on two types of phosphorus resins (with phosphinic or phosphoric functional groups), it is likely that resins with a phosphonic group should not be particularly heat sensitive too. This is confirmed by a uranium extraction study with Amberlite IRC747 resin between 40 °C and 60 °C [14].

The state-of-the-art literature has shown that commercial resins containing aminophosphonic groups (pattern (a) in Figure 1), such as the reference CH-93 Tension provided by Thermax, Lewatit TP-260, Purolite S940 and Amberlite IRC747, exhibit high REE complexing ability. In order to test the phosphonic functional group without amine, the Diphonix resin manufactured by EiChrom Industries (pattern (c) in Figure 1) has also been selected. The Actinide-Resin B (pattern (d)) has also been tested because of its very high complexing power [15]. For the Lewatit VP OC 1026 resin (pattern (b) in Figure 1), the functional group (D2EHPA) is impregnated (by adsorption) in the crosslinked polystyrene divinylbenzene matrix. For the other resins the functional groups are covalently bonded to the cross-linked polystyrene divinylbenzene matrix.

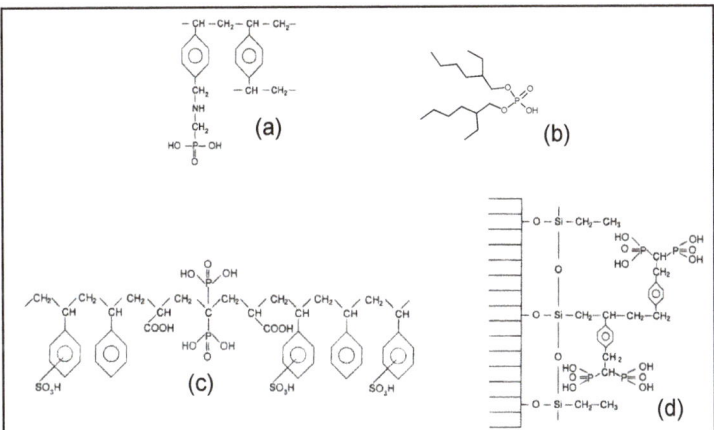

Figure 1. Chemical structure of functional groups present in various commercial resins potentially effective for extracting rare earth elements (REE) in a phosphoric medium. (**a**) Tulsion CH-93, Purolite S940, Amberlite IRC-747 and Lewatit TP-260, (**b**) Lewatit VP OC 1026, (**c**) Diphonix, (**d**) Actinide-CU (reproduced from [15], with permission of American Chemical Society, 1998).

3. Materials and Methods

Tulsion CH-93 was supplied by Thermax (Pune, India), Diphonix and Actinide-CU were delivered by EiChrom Industries (Bruz, France), Amberlite IRC 747, and Purolite S940 were from Dow France SAS (Chauny, France) and Purolite International (Paris, France), Lanxess Energizing Chemistry (Courbevoie, France), supplied Lewatit TP-260 and VP OC 1026. These resins have been preconditioned in deionized (DI) water. They are all weak acid ion exchangers and are supplied as powders composed of spherical beads. Their main characteristics are reported in Table 1.

It is worth mentioning that preliminary conditioning of Na^+-resins under H^+ ionic form has been proved to have no measurable impact on the extraction efficiency of the resins for all the REE and impurities present in the tested phosphoric acid solutions. The high acidity of the phosphoric acid solutions compared to the low concentrations in impurities favors the complete H^+/Na^+ exchange in the resins before REE extraction. The variation in the concentration of H^+ in the bulk of the acid solution due to either the REE adsorption or the Na^+ exchange is negligible when compared to its initial concentration in the phosphoric acid solution.

3.1. Physicochemical Properties of Resins

Specific surface areas of resins were determined by measuring the nitrogen adsorption/desorption isotherm. The latter is represented in the form of a graph giving the adsorbed quantity per gram of adsorbent as a function of the equilibrium relative pressure (ratio of the equilibrium pressure of the adsorbable gas and its saturation vapor pressure). The plot of this isotherm allows, in the zone of relative pressure between 0.3 and 0.8, to determine specific surface areas using the model Brunauer, Emmett, Teller (BET). This isotherm also gives pore size distribution with the model Barret Joyner, Halenda (BJH). Table 1 gathers the measurements of specific surface areas and pore sizes according to the BET and BJH models respectively.

Table 1. Main characteristics of the resins tested in this study.

Functional Group	Name of Resin	Ionic Form	Bead Size from Suppliers (min 95%) (mm)	Specific Surface Area (BET) m²/g	Pore Size (BJH Desorption) (nm)	Densities
Amino phosphonic	Tulsion CH-93	Na^+	0.3–1.2	28	40–50	0.612
	Purolite S940	Na^+	0.43–0.85	21	40–50	0.469
	Amberlite IRC-747	Na^+	0.52–0.66	21	40–50	0.530
	Lewatit TP-260	Na^+	0.58–0.68	15	40–50	0.726
DIPEX Extractant	Actinide Resin-B	Na^+	0.1–0.15	76	20	0.416
D2EHPA	Lewatit VP OC 1026	H^+	0.31–1.6	5.5	50	0.607
Monophosphonic + sulfonic	Monophos	H^+	0.3–1.2	0.66	30–40	-
Diphosphonic + sulfonic	Diphonix	H^+	0.3–1.2	0.14	Not measurable	-

A large specific surface area (i.e., >15 m²/g) and pore diameters between 2 and 50 nm are sought in order, on the one hand, to maximize the ion exchange capacity and, on the other hand, to promote the diffusion of solute in the complexation sites located in the core of the material. Resins with aminophosphonic function (CH-93, S940, IRC-747, TP-260) have specific surface areas between 16 and 28 m²/g. Actinide Resin B has a surface area two to three times higher while VP OC 1026 leads to a lower value of 5.5 m²/g. All the resins containing sulfonic groups have very low specific surface areas (<1 m²/g), which could limit the accessibility to the pores. Overall, the resins not containing sulfonic groups show specific surface areas and pore sizes that meet targeted characteristics of a mesoporous resin. Nevertheless, BET and BJH measurements are not sufficient to compare the resins in terms of exchange capacity and specific experiment have been conducted in Section 4.3.

3.2. Composition of Aqueous Solutions

Table 2 gives the composition in REE of the solutions used in this study. J0 and J1 are genuine phosphoric acid solutions produced by OCP SA (Jorf Lasfar El Jadida, Morocco). Both solutions have been analyzed by Inductively Coupled Plasma-mass Spectrometry (ICP-MS) instrument (ThermoFischer Scientific, Villebon-sur-Yvette, France) (error < 5%). From these measurements synthetic solutions (called REH and REOCP, composition in Table 2) have been made by solubilizing the major REE (>mg/L in bold in the J1 column) at a higher concentration than in the genuine samples. This procedure enables the use of ICP-AES (Horiba Jobin Yvon SAS, Longjumeau, France) measurements with a sufficient analytical sensitivity (error < 5%) while minimizing the measurement time. In order to be representative of the genuine solution, the REE were solubilized in 4.2 mol/L phosphoric acid, from their metallic form for La, Nd, Gd, Dy, Er, Yb and from their oxide form for Sc and Y. REE are solubilized as trivalent cations. The difference in REH and REOCP solutions lies in the presence of major interfering metallic impurities in the REOCP solution. In order to be similar to a genuine solution, vanadium, magnesium, and zinc were solubilized from their metallic form in 4.2 mol/L phosphoric acid, while the iron was prepared from its hydrate phosphate solid form. Aluminum was solubilized from its hydroxide form and calcium from carbonate. The error in the measurement of these impurities is ±5% in the whole range of concentrations whatever the technique (ICP-MS or ICP-AES). J0 has been used for kinetic studies while J1, JSYNT, REH and REOCP have been used for extraction yields measurement in batch experiments

Ion exchange resins generally have greater selectivity for ions with increasing valence or charge. Among the ions with the same charge, higher affinity is observed for ions with higher atomic number Z and the extraction efficiency decreases with increasing REE ionic radius IR [9]. It is worth mentioning that the chosen major REE in the synthetic solution have ionic radii that cover the whole range of the actual REE ionic radii (see Table 2).

Table 2. Composition in REE (mg/L) and in major interfering impurities (*in italic, g/L*) of the phosphoric acid solutions (J0-J1-JSYNT, REH, REOCP) used in the experiments with corresponding ionic radii IR (data from [16])and ratio Z/IR where Z is the atomic number of the REE.

C_M (mg/L) or (g/L)	J0	J1	REH REOCP	JSYNT	Ionic Radius [16] (A)	Z	Z/IR (A^{-1})
Sc	6.5	10	72	6.7	0.68	21	31
Y	24.6	37	72	6.8	0.89	39	44
La	4.5	8.8	53	7.5	1.06	57	54
Pr	1.2	1.3	-	-	1.01	59	58
Nd	3.1	5.8	54	5.0	1.00	60	60
Sm	0.9	1.5	-	-	0.96	62	65
Eu	0.3	0.4	-	-	0.95	63	66
Gd	1.7	2.7	72	6.4	0.94	64	68
Tb	0.4	0.4	-	-	0.92	65	71
Dy	1.9	3.0	74	6.8	0.91	66	73
Ho	0.7	0.8	-	-	0.89	67	75
Er	1.9	2.8	73	6.7	0.88	68	77
Tm	0.5	0.5	-	-	0.87	69	79
Yb	2.8	4.1	74	6.8	0.86	70	81
Lu	0.8	0.8	-	-	0.85	71	84
Al	*0.9*	*1.1*	*0.8*		-	-	-
Ca	*0.8*	*0.8*	*1.0*		-	-	-
Fe	*1.7*	*1.4*	*1.4*	*1.3*	-	-	-
Mg	*3.9*	*4.3*	*4.8*		-	-	-
V	*0.12*	*0.11*	*0.1*		-	-	-
Zn	*0.22*	*0.21*	*0.30*		-	-	-
[H_3PO_4] (mol/L)	3.9	3.8	4.2	4.2	-	-	-

3.3. Experimental Protocol for Batch Tests (Extraction and Elution)

Sorption or elution experiments are carried out in 15 mL or 2 mL tubes arranged horizontally on a stainless steel support with a double jacket to keep the temperature constant, with 1 or 10 mL of aqueous phosphoric acid solutions and 50 or 100 mg of resin previously dried overnight in an oven at 60 °C. Tubes are stirred at 50 °C (respectively 25 °C for elution studies) in Vibrax for 3 h or in Ecotron for 24 h.

Once stirring is complete, the aqueous solutions are filtered under vacuum using a 0.45 µm filter. The tube and filter are washed with DI water to remove the impregnated aqueous extractant solution and then the powder is dried in an oven at 60 °C overnight.

Results obtained after stirring for 3 or 24 h were similar, thus indicating that the sorption or elution equilibrium has been reached for all experiments.

3.4. ICP-AES Analysis Protocol

ICP-AES is a very sensitive elemental analysis technique used to quantify the total amount of elements in aqueous solution. The analysis is carried out on samples diluted in 0.42 mol/L phosphoric acid. The device used is an Ultima-2 brand Horiba Jobin Yvon SAS.

The wavelengths chosen for the ICP-AES are as follows (in nm): Al 396.152—Ca 393.366—Ce 413.765—Dy 394.469—Er 390.631—Fe 259.837—Gd 301.014—La 399.575—Mg 518.362—Nd 406.109—Sc 337.215—V 311.838—Y 377.433—Yb 328.937—Zn 206.200.

Al, Ca, Fe, Mg, V, and Zn are calibrated in the range 1 to 10 mg /L (0—1—3—5—10 mg/ L). Ce, Dy, Er, Gd, La, Nd, Sc, Y, and Yb are calibrated from 0.1 to 1 mg/L (0—0.1—0.3—0.5—1 mg/L).

The error in concentration measurement has been estimated to be ±5% taking into account the dilution.

3.5. Mineralization Protocol

The mineralization operation consists in dissolving solids or decomposing organic material to obtain an aqueous solution for elemental analyses by ICP-AES. The device used is an Ethos One from Milestone. It comprises an oven, inside of which is placed a rotor comprising 10 segments each containing a 100 mL reactor. The microwave oven serves to improve the dissolution kinetics of solids or decomposition of organic matter. The reference reactor contains a thermowell in which the thermocouple plunges to pilot the reaction. A minimum of 8 mL of solution with at least 4 mL of HNO_3 69 wt.% (15.4 mol/L) and 4 mL of hydrogen peroxide 35 wt.% are introduced into the reactors with 50 mg of resin. Mineralization consists of three main steps:

1. Temperature increase: gradient 10 °C/min for 20 min up to 200 °C and 100 bars at maximum power (1000 W).
2. Temperature step: 10 min.
3. Cooling: 30 min during which a forced convection with a fan occurs in the oven.

The reactor is then rinsed with 20 mL of DI water recovered in a volumetric flask. After dilution in H_3PO_4 0.42 mol/L, mineralization solutions are analyzed by ICP-AES.

3.6. Sorption Yields Y_M and R_M of a Cation M

Sorption yield corresponds to the amount of resin-fixed cations relative to the amount initially present in solution. Two calculation modes can be performed:

- By balancing the concentrations (in mg/L) in aqueous phase before (C_{M-ini}) and after extraction (C_{M-end}):

$$Y_M \text{ (extraction)} = (C_{M-ini} - C_{M-end})/C_{M-ini}$$

- By balancing the masses of cation M in the aqueous phase before extraction and fixed onto the resin. This last mass is determined by the concentration of the mineralization solution, $C_{M-miner}$ (mg/L), and by the knowledge of the total masses and volumes used for the extraction (m_{ext} (mg), V_{ext}(L)) and the mineralization (m_{miner} (mg), V_{miner} (L)):

$$R_M \text{ (extraction)} = (C_{M-miner} V_{miner})/(C_{M-ini} V_{ext})(m_{ext}/m_{miner})$$

The errors on Y_M and R_M are estimated to be ±15%.

3.7. Mass of Adsorbed Cation Per Gram of Dry Resin (X_M)

Two calculation modes can be performed:

- Taking into account initial and post-extraction concentrations in the aqueous phase:

$$X_M = (C_{M-ini} - C_{M-end}) V_{ext}/m_{ext}$$

The error on X_M is estimated to be ±15%.

- By direct mineralization of the resin trapping the cation:

$$X_{M\text{-fixed}} = C_{M\text{-miner}} V_{miner} / m_{miner}$$

The error on X_M based on that calculation is estimated to be ±10%. It is worth mentioning that X_M is often referred to as the resin adsorption capacity for cation M.

3.8. Experimental Protocol for the Kinetic Studies

These studies have been performed using lab-scale columns in which a resin bed is put. Before starting the REE extraction from phosphoric acid, the resin was conditioned according to the following steps:

- Washing the resin with demineralized water.
- Conditioning the resin under H^+ ionic form using sulfuric acid.
- Washing the resin to remove the remaining sulfuric acid.

Then the real phosphoric acid is continuously passed through the resin bed whose volume (BV) is 50 mL under a constant flowrate of 5 mL/min (0.1 BV/min) so that the residence time inside the resin bed is 10 min. Every 10 min., a sample of the phosphoric acid is collected at the outlet of the column and analyzed by ICP-AES (error 5%). This sampling has been done over 90 min. (i.e., 9 BV).

3.9. Mass of Eluted Cations Per Gram of Dry Resin ($X_{M\text{-eluted}}$)

The calculation takes into account only the mass of resin used for the elution $m_{elution}$, the aqueous elution volume $V_{elution}$ (L) and the concentration of the cation in aqueous phase after elution $C_{M\text{-elution}}$ (mg/L)

$$X_{M\text{ eluted}} = C_{M\text{-elution}} V_{elution} / m_{elution}$$

The error on X_M based on that calculation is estimated to be ±10%.

4. Results and Discussion

4.1. Extraction of REE from Synthetic Solutions REH and REOCP

In order to estimate the extraction efficiency of REE by resins, experiments were carried out following the protocol reported in Section 3.3 with 100 mg of resins contacted by 10 mL of REH or REOCP solution. Some of the resins were destroyed by mineralization and the resulting aqueous phases analyzed by ICP-AES. Sorption yields at 50 °C (respectively quantities of REE fixed on one gram dry resin) are reported in Figure 2 (resp. Figure 3) as functions of Z/IR and in Figure 4 as functions of IR.

Within the margin of error (10% or 15%), results deduced from the mineralization are consistent with calculations from aqueous phases, thus validating the mineralization protocol and the good accuracy of the ICP-AES measurement. The deviation between both extraction yields is generally less than 15%, except for the VP OC 1026 resin, which is possibly due to the use of a hydrated resin, hence an overestimation of the mineralized solid mass. In the following, only aqueous concentrations have been considered.

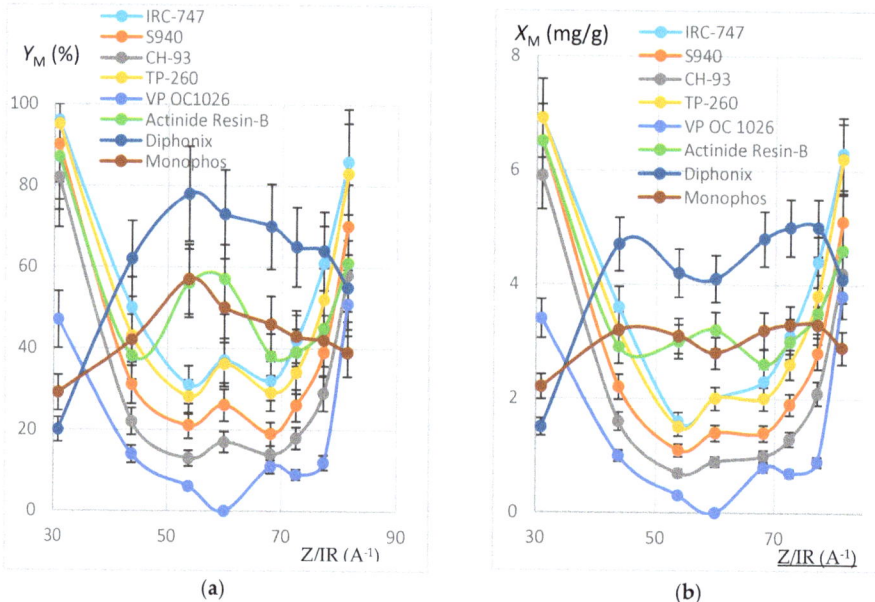

Figure 2. (a) Plot of Y_M (%) as a function of Z/IR (A^{-1}). REH solution. Error bars ±15%. (b) Plot of X_M (mg/g) as a function of Z/IR (A^{-1}). REH solution. Error bars ±10%.

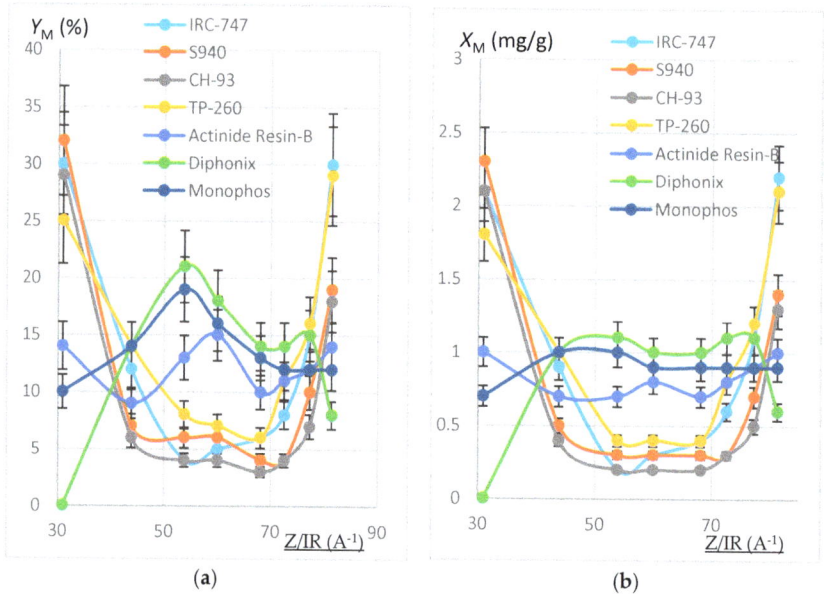

Figure 3. (a) Plot of Y_M (%) as a function of Z/IR (A^{-1}). REOCP solution. Error bars ±15%. (b) Plot of X_M (mg/g) as a function of Z/IR (A^{-1}). REOCP solution. Error bars ±10%.

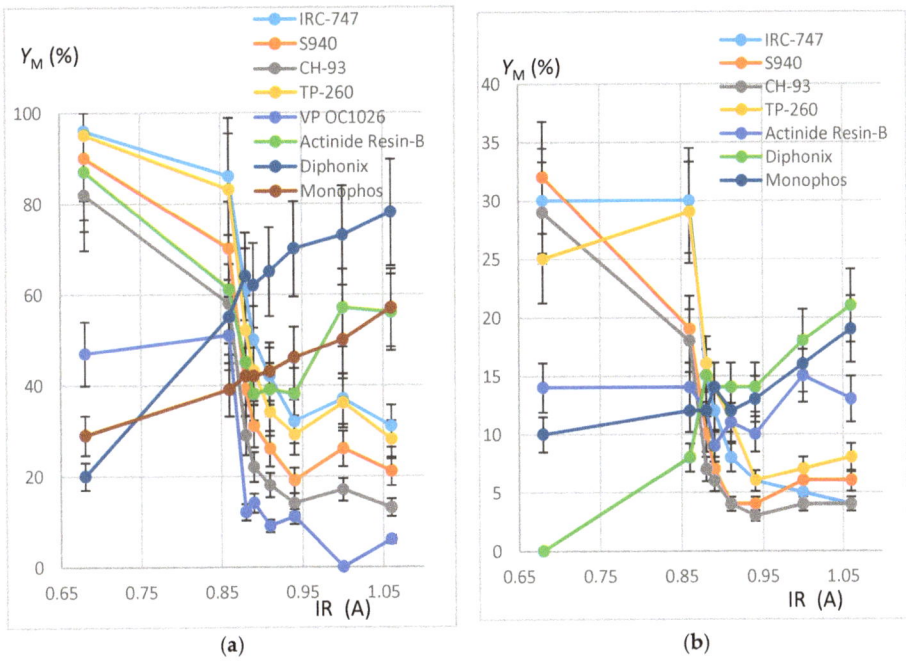

Figure 4. (a) Plot of Y_M (%) as a function of IR (A). REH solution. Error bars ±15%. (b) Plot of Y_M (%) as a function of IR (A). REOCP solution. Error bars ±15%.

It can be noticed that each resin maintains the same behavior towards the REE extraction when the impurities are present or not in the synthetic REH and REOCP solutions:

- The behavior of the aminophosphonic resins IRC-747, S940, CH-93, and TP-260 is similar: extraction decreases when Z/IR increases from the scandium up to a plateau constituted by the group of La, Nd, Gd, and then increases with Z/IR from Dy to Yb. It is worth noting that the efficiency of extraction (see Figure 4) decreases with IR similarly to that observed with CH-93 resins [12,13,17] or with phosphorus extractants [18], for example dialkylphosphinic acid CYANEX 272 [19] or di(2-ethylhexyl)phosphoric acid D2EHPA [20]).
- Without impurities in the phosphoric acid solution, the VP-OC-1026 resin on which the D2EHPA is adsorbed behaves similarly with the aminophosphonic resins but with a lower sorption efficiency.
- Without impurities, the extraction efficiency of Actinide Resin-B is not clearly related to Z/IR or to IR (see Figure 4). The presence of impurities flattens the extraction efficiency, which becomes nearly constant between 10% and 15% for all of the REE.
- The presence of a sulfonic group together with an alkylphosphonic function in the Diphonix and Monophos resins results in an inverted behavior compared to the above resins: the extraction efficiency increases as Z/IR increases from the scandium to the neodymium and then decreases for the highest Z/IR although it continuously increases with IR (see Figures 2–4). Despite their low specific surface area compared to the tested aminophosphonic resins (see Table 1) the Diphonix and Monophos resins extract REE in the same order of magnitude (from 10 to 20% except for scandium and ytterbium for the Diphonix). It is then probable that the number of accessible sites of complexing functions per gram of resin is higher for Diphonix and Monophos than for the other resins.

In a more general manner, these results highlight the influence of the atomic number on the extraction affinity of the resins towards the REE and of the particular behavior of Diphonix/Monophos resins compared to the others.

Finally, the total masses of extracted REE and impurities are shown in Figure 5 for each resin together with the mass of the extracted iron. Within the margin of errors, all the resins have loaded the same mass of REE (about 7 mg/g), except for CH-93 whose loaded mass is smaller than the others. With aminophosphonic resins (IRC-747, TP-260, S940 and CH-93) iron is the most extracted element that extraction corresponding to 62–84% of all the moles of cations sorbed onto these resins. The other resins extract less iron but much more magnesium, which also limits the sorption of REE. Overall, IRC-747, TP-260, and Diphonix/Monophos prove to be the most effective for extracting heavy REE and scandium: approximately 0.07–0.08 millimole of REE per gram of dry resin is fixed. Compared to the total mass extracted in REE of about 0.3 millimole per gram of resins without impurities the competition with iron, aluminum, and magnesium reduces by a factor of four the total mass of REE extracted from a representative OCP phosphoric acid.

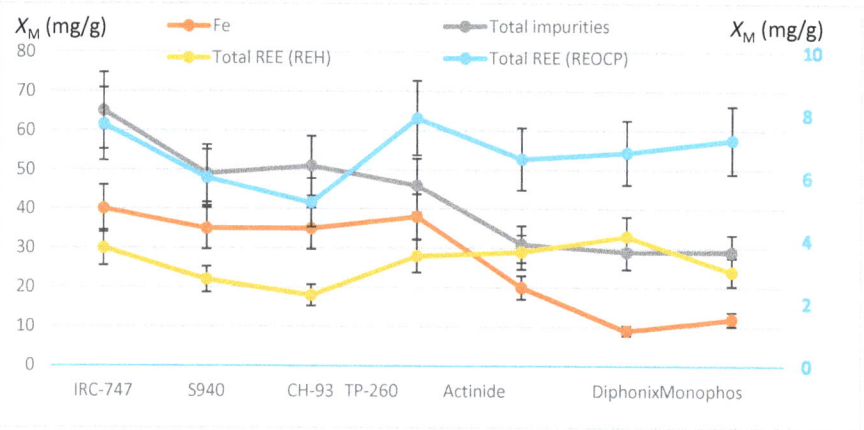

Figure 5. Comparison of total REE uptakes in the presence of impurities (REOCP synthetic solution, right y-axis) or not (REH synthetic solution, left y-axis). Plots of total impurities and iron uptakes (left y-axis) for the different resins. Error bars ±15%.

4.2. Extraction of REE from JSYNT and J1 OCP Phosphoric Acid Solutions

In order to quantify the lone interference of iron in the REE extraction the synthetic JSYNT solution has been made with concentrations of iron and REE close to those of the genuine J1 solution (see Table 2). From the preceding results, IRC-747 and TP-260 resins have been chosen due to their good affinities towards REE. They have been put in 2 mL tubes with 50 mg of resins (see protocol in Section 3.3). Sorption yields Y_M and the masses of elements X_M trapped onto one gram of dry resin are illustrated in Figure 6a,b respectively. X_M values have been calculated based on the difference in aqueous concentrations before and after extraction.

It is worth mentioning that the error in Y_M is about 15% so that the deviations between J1 and JSYNT for IRC 747 are significant while the extraction yields for IRC 747 and TP260 are within the margin of error. These remarks hold for X_M for which the error is estimated to be around 10%.

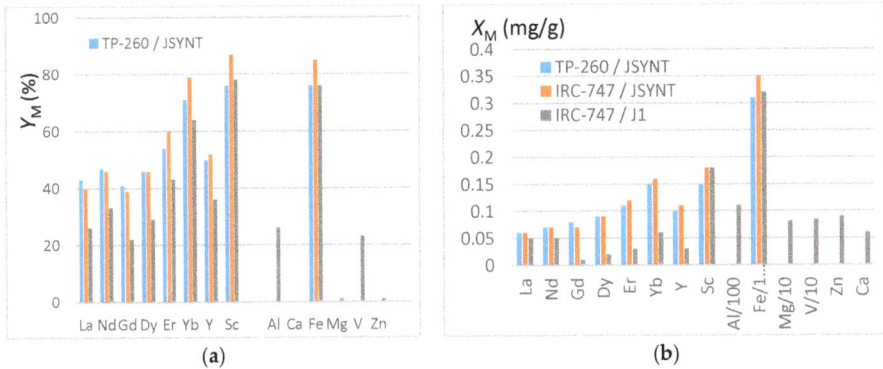

Figure 6. (a) Extraction yields Y_M from JSYNT and J1 solutions. (b) Masses X_M (mg/g) of REE and impurities fixed in one gram of dry resin. Note that the fixed masses of impurities are divided by 100 (Al, Fe) or by 10 (Mg, V).

As previously noticed, iron is extracted with the same efficiency than scandium, the most extracted REE. The extraction yields are higher here because of a ratio of resin mass to aqueous volume higher than previously used (50 mg/1.5 mL instead of 100 mg/10 mL). Aluminum and vanadium are extracted with the same efficiency as light REE (La, Nd and Dy), which leads to a significant amount of these impurities trapped in the resins. It is worth mentioning that these efficiencies are coherent with those reported by Rhadika et al. [12] for the aminophophonic acid resin Tulson-CH 93. However, when expressed as X_M, these results highlight the cumulative deleterious influence of metallic impurities (notably Fe and Al) on the REE extraction from genuine phosphoric acid. As an illustration, IRC-747 and TP-260 extract 0.009 millimole of total REE per gram of dry resin compared to about 0.6 millimole of iron from J1.

4.3. Adsorption Isotherm of Er

Erbium is one of the heavy REE contained in OCP phosphoric acid that is readily extracted by IRC-747 or TP-260 resins. For estimating the ion exchange capacity of the resins, erbium extraction tests were carried out in 15 mL tubes with 100 mg of resins previously dried overnight in an oven at 60 °C, and 10 mL of aqueous solution containing variable concentrations of erbium dissolved in pure phosphoric acid at 4.2 mol/L (see Section 3.3).

The linearized Langmuir adsorption isotherm is expressed by the following equation [21]:

$$\frac{C_{eq}}{X} = \frac{1}{bX_m} + \frac{C_{eq}}{X_m}$$

where C_{eq} is the concentration of erbium at equilibrium in aqueous phase (mg/L), X is the amount of Er extracted by IRC-747 or TP-260 (mg/g), b is the Langmuir constant (L/mg), and X_m is the maximum adsorption capacity (mg/g). A very small value of Langmuir constant could be correlated to favorable adsorption thermodynamics. Figure 7 proves that the experimental adsorption isotherms of Er using the above resins fit very well the linearized Langmuir adsorption isotherm. The experimental capacity X_m and b are then obtained from slope and intercept of the straight lines shown in Figure 3. According to this model, the absorption of metal ions occurs on a homogeneous surface by monolayer adsorption and there is no interaction between the sorbed species.

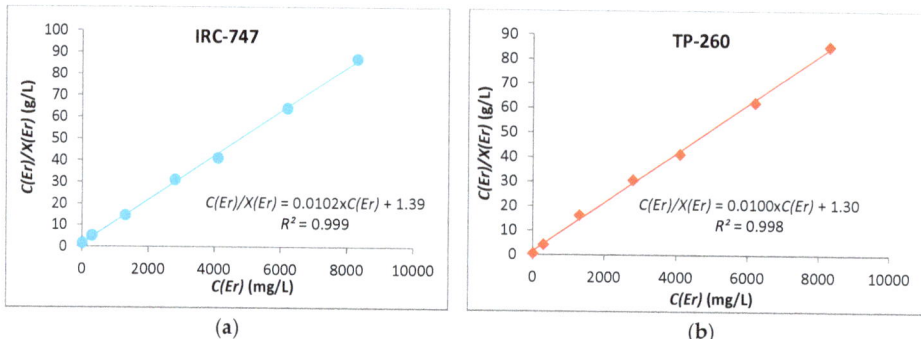

Figure 7. Langmuir adsorption isotherm for Er(III) with IRC-747 (**a**) and TP-260 (**b**).

4.4. Extraction Kinetics with IRC-747 and TP-260

In this section, the REE extraction kinetics from the genuine J0 solution (see Table 2) are presented in Figures 8 and 9, corresponding respectively to the IRC-747 and TP-260 resins. The experimental protocol is mentioned in Section 3.8. The concentrations of C_M in REE have been measured by ICP-AES at the outlet of the columns for each bed volume BV after having passed the real phosphoric acid through the resin (see Figures 8a and 9a). The corresponding extraction yields expressed as Y_M are reported on Figures 8b and 9b.

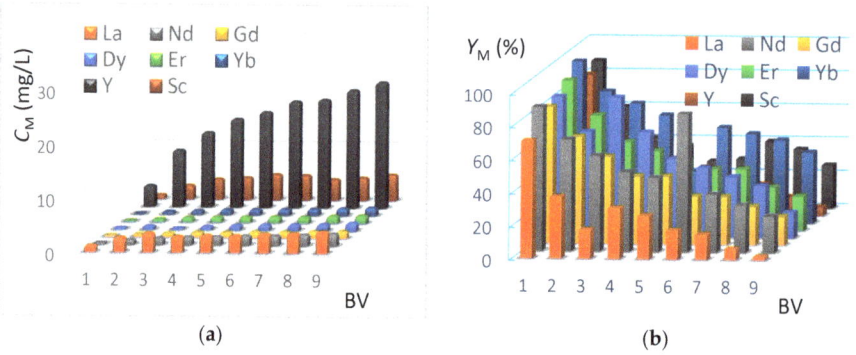

Figure 8. (**a**) Concentrations of REE at the outlet of the TP-260 column as a function of bed volume (BV = 50 mL, flowrate = 5 mL/min). Genuine J0 solution. (**b**) Extraction yields of REE at the outlet of the TP-260 column as a function of bed volume (BV = 50 mL, flowrate = 5 mL/min). Genuine J0 solution.

The extraction results using TP-260 resin show that more than 84% recovery of the total REE contained in phosphoric acid is achieved for the first bed volume. Ytterbium, erbium, scandium, and yttrium present the highest yield of extraction, while lanthanum represents the lowest yield of extraction. This is consistent with batch tests done before (see Section 4.2). Concerning the second bed volume passing through the resin, we can notice that the extraction efficiency is 30% lower for the REE, and at the ninth bed volume, the extraction yield is 13%, which shows that the resin is almost saturated.

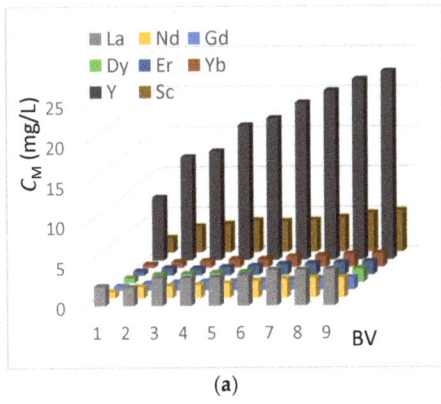
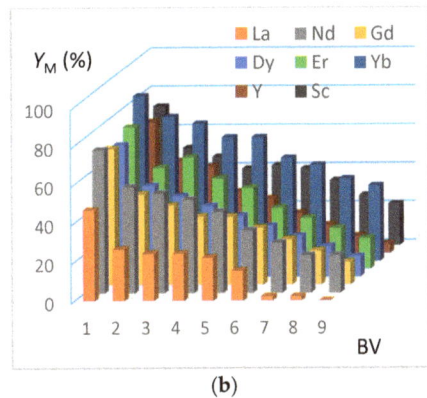

Figure 9. (a) Concentrations of REE at the outlet of the IRC-747 column as a function of the number of bed volume passed (BV = 50 mL, flowrate = 5 mL/min). (b) Extraction yields of REE at the outlet of the IRC-747 column as a function of the number of bed volume passed (BV = 50 mL, flowrate = 5 mL/min).

As in the case of TP-260 resin, the extraction efficiency is the highest for ytterbium, neodymium, erbium, and yttrium but values are somewhat lower: around 70% are recovered with IRC-747 for the first bed volume passing through the resin, instead of 84% for TP-260. For both resins, lanthanum is the less extracted element. The results also show that the resin is saturated with lanthanum within the 8th bed volume.

A comparison between the kinetics of extraction of TP-260 and IRC-747 is shown in Figure 10a where F denotes the total REE normalized concentration at time t expressed as the ratio of the fixed total REE concentration at the outlet of the resin to the initial total REE concentration at the inlet of the resin. F is the fractional attainment of equilibrium adsorption at time t.

The adsorption onto ion exchange resins must be considered as a liquid–solid phase reaction, which includes several steps [21,22]:

- The diffusion of ions from the solution to the resin surface whose rate R can be described by:

$$R = -\ln(1 - F) = kt \tag{1}$$

- The diffusion of ions within the solid resin whose rate R can be described by:

$$R = -\ln(1 - F^2) = kt \tag{2}$$

- And the chemical reaction between ions and functional groups (of the resin).

When the adsorption of metal ion involves mass transfer accompanied by chemical reaction (also referred to as the moving boundary model) the rate equation R is given by:

$$R = 3 - 3(1 - F)^{2/3} - 2F = kt \tag{3}$$

All these models have been compared against experimental data; it appears that the best fit is obtained when the metal uptake is controlled by the moving boundary particle diffusion model during the first 80 min (8 BV) for TP-260 and 70 min for IRC-747 (Figure 10b).

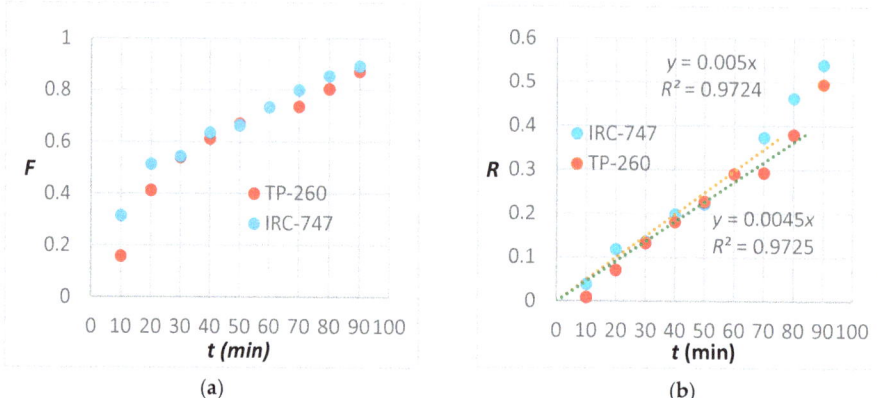

Figure 10. (a) Plot of the total REE adsorption on Lewatit TP260 and Amberlite IRC-747 resins. (b) Plot of moving boundary particle diffusion model for the total REE adsorption on Lewatit TP260 and Amberlite IRC-747 resins.

In Equation (3), k is given by the expression:

$$k = \frac{6 D_r C_{M-ini}}{a\, r_0^2 C^s_{M-ini}}$$

where D_r represents the diffusion coefficient in the resin bead (m^2/s), r_0 is the particle radius (m), C^s_{M-ini} is the concentration (mg/L) of solid reactant at the bead's unreacted core and a is the stoichiometric coefficient.

4.5. Elution Conditions in Carbonate or Sulfuric Media

The objective here is to evaluate different aqueous phase compositions for the quantitative and if possible selective elution of extracted cations. In order to be close to an industrial elution process, the protocol has consisted first in washing and drying the resin powder overnight at 60 °C after filtration (0.2 µm) and then in contacting the resin containing cations with aqueous elution solution at 25 °C. From economic considerations, this elution solution must be either a by-product of the phosphoric acid production or sufficiently cheap. Carbonate and sulfuric acid solutions have thus been considered here instead of concentrated hydrochloric acid for example.

This elution was carried out in 15 mL tubes containing 50 mg of loaded resins with REOCP solution. These resins have been previously dried overnight in an oven at 60 °C and put in contact during 3 h with 10 mL of stirred solutions of sodium carbonate 1 mol/L or sulfuric acid 9 and 18 mol/L. These tubes were kept at a constant temperature of 25 °C. All aqueous phases were analyzed by ICP-AES. The results are expressed as a yield of elution (i.e., the ratio of actual eluted mass of cations to the mass of cations initially fixed per gram of resin) and are illustrated in Table 3 following the two modes of calculation (Min: resin mineralization. Aq: aqueous phases). The deviation between both methods is within the margins of error of these protocols.

From that table it appears that even at low concentrations (i.e., <10%), the elution efficiency of iron makes the concentration of this impurity in the eluting solution always bigger by an order of magnitude than the concentration of eluted REE. Thus, from an engineering point of view the simultaneous extraction and separation of iron and REE from the high concentrated OCP phosphoric acid appear to be not feasible with IRC 747 and TP-260 resins and low-cost carbonates or sulfuric acid eluting solutions.

Table 3. Elution yields in sodium carbonate or sulfuric acid of the elements from resins previously contacted with the REOCP solution. Aq: calculations based on aqueous phases. Min: calculations based on solid mineralization (V_{Aq} = 10 mL, m_{resin} = 50 mg, stirring 3 h, T = 25 °C).

Name of Resin	IRC-747		IRC-747		TP-260		TP-260
Eluting Solution	H_2SO_4 9M		Na_2CO_3 1M		Na_2CO_3 1M		H_2SO_4 18M
Y_M, R_M Elution (%)	Aq	Min	Aq	Min	Aq	Min	Aq
La	39	34	25	30	20	15	72
Nd	36	31	47	47	45	41	68
Gd	44	36	60	46	51	46	79
Dy	36	34	58	51	57	43	80
Er	32	28	68	55	70	56	79
Yb	29	26	66	56	63	53	78
Y	30	29	65	45	61	43	78
Sc	0	0	50	43	36	32	34
Al	9	9	30	25	31	29	0
Ca	48	68	65	-	75	-	3
Fe	4	3	9	7	11	8	8
Mg	80	15	14	2	19	18	16
V	35	26	100	81	106	92	0
Zn	83	-	-	-	-	-	79

5. Conclusions

Among commercially available chelating ion exchange resins, IRC-747 and TP-260 appears to be particularly interesting for the extraction of REE from the high concentrated phosphoric acid (about 4 mol/L) produced by OCP due to their highest affinity for the metallic cations. Even if the sorption mechanisms are not understood, the influence of the atomic number and the ionic radius of the REE on the sorption affinity result in extraction yields that range from 20 to 60% and even about 80% for scandium. The adsorption of erbium (III) from pure 4 mol/L phosphoric acid can be described by the Langmuir isotherm and leads to a maximum adsorption capacity estimated to be 1.8 meq/g for both resins. This value is similar to the total mass of iron fixed in the resins when put in contact with the OCP phosphoric acid. The capacity of the TP-260 and IRC-747 resins used for the extraction of REE from concentrated phosphoric acid solutions are thus far larger from other resins tested in the literature. However, the uptake of competitive pollutants, among which iron and aluminum appear to be the most interfering, reduces the REE adsorption by a factor of 4. The elution procedure using high concentrated sulfuric acid enables the recovery of a large fraction of REE (around 80% except for scandium at around 30%); however, even if the elution yield for iron is much lower (<10%) the resulting elution solution needs to be further processed for the recovery of a pure REE concentrate. Furthermore, the build-up in iron inside the resin bed will also keep growing during the extraction/elution cycles. It is then mandatory to find another eluting system, for example based on a more expensive but selective complexing agent for REE such as diglycolamide.

Also, the kinetics of the total REE uptake by IRC 747 and TP 260 from a genuine phosphoric acid solution have been tested against different model expressions. The data appear to be best explained by the moving boundary particle diffusion model during the first 7th or 8th bed volume passed. Other operating conditions (flowrate, mass of resin ...) should be required for validating such a behavior.

Finally, the preliminary OCP proprietary treatment has separated uranium and thorium from the produced phosphoric acid but the presence of other metallic impurities at relatively high concentration are likely to limit the techno-economical interest of the chelating ion exchange processes. Complementary studies could be conducted in order to check whether iron (and aluminum) could be kinetically separated from REE, even if this solution seems to be difficult to implement at an industrial scale. Recent developments in solvent extraction (SX) in high concentrated phosphoric acid could overcome this obstacle due to the high separation factors between REE and iron(III) [23]. However, this solution has still to be validated at larger scales.

Author Contributions: Conceptualization, X.H., V.B., A.O., H.M. and D.D.; Methodology, X.H., V.B., P.D.N., A.O., D.D. and F.C.; Validation, X.H., P.D.N., A.O., H.M. and D.D.; Formal Analysis, V.B. and X.H. Investigation, X.H., P.D.N., F.C., A.O., H.M. and D.D.; Resources, X.H., P.D.N., V.B., A.O., H.M. and D.D.; Writing-Original Draft Preparation, X.H., P.D.N., A.O., H.M. and D.D.; Writing-Review & Editing, V.B. and X.H. Visualization, V.B., X.H. and A.O. Supervision, V.B., A.O., D.D.; Project Administration, V.B, A.O., D.D.; Funding Acquisition, D.D., H.M., A.O. and V.B.

Funding: This research received no external funding.

Acknowledgments: The authors want to acknowledge OCP for financial support and an anonymous reviewer for their valuable comments. The authors also would like to acknowledge the support of national network Promethee.

Conflicts of Interest: The authors declare no conflict of interest.

References

1. Soldenhoff, K.; Tran, M.T.; Griffith, C. Recovery of Uranium from Phosphoric by Ion Exchange. In Proceedings of the 2009 IAEA Technical Meeting—Uranium from Unconventional Resources, Vienna, Austria, 22–26 June 2009.
2. Kim, H.; Eggert, R.G.; Carlsen, B.W.; Dixon, B.W. Potential uranium supply from phosphoric acid A U.S. analysis comparing solvent extraction and Ion exchange recovery. *Resour. Policy* **2016**, *49*, 222–231. [CrossRef]
3. Michel, P. *Les Techniques de L'industrie Minérale*; Société de L'industrie Minérale: Paris, France, 2006; Volume 32, pp. 95–102.
4. Cote, D.B.G.; Mokhtari, H.; Courtaud, B.; Moyer, B.A.; Chagnes, A. Recovery of Uranium from Wet Phosphoric Acid by Solvent Extraction Processes. *Chem. Rev.* **2014**, *114*, 12002–12023.
5. Volkman, Y. *Recovery of Uranium from Phosphoric Acid by Ion Exchange*; Report IAEA-TECDOC-533; IAEA: Vienna, Austria, 1987; pp. 59–68.
6. Mashkovtsev, M.; Botalov, M.; Smyshlyaev, D.; Pajarre, R.; Kangas, P.; Rychkov, V.; Koukkari, P. Pilot-scale recovery of rare earths and scandium from phosphogypsum and uranium leachates. In Proceedings of the 2016 Mineral Engineering Conference, Swieradow-Zdroj, Poland, 25–28 September 2016.
7. Page, M.J.; Soldenhoff, K.; Ogden, M.D. Comparative Study of the Application of Chelating Resins for Rare Earth Recovery. *Hydrometallurgy* **2017**, *169*, 275–281. [CrossRef]
8. Helfferich, F. *Ion Exchange*; McGraw-Hill: New York, NY, USA, 1962.
9. Reddy, B.R.; Kumar, J.R. Rare Earths Extraction, Separation, and Recovery from Phosphoric Acid Media. *Solvent Extr. Ion Exch.* **2016**, *34*, 226–240. [CrossRef]
10. Quinn, J.E.; Soldenhoff, K.H.; Stevens, G.W.; Lengkeek, N.A. Solvent extraction of rare earth elements using phosphonic/phosphinic acid mixtures. *Hydrometallurgy* **2015**, *157*, 298–305. [CrossRef]
11. Nesterenko, P.N.; Zhukova, O.S.; Shpigun, O.A.; Jones, P. Synthesis and Ion-Exchange Properties of Silica Chemically Modified with Aminophosphonic Acid. *J. Chromatogr. A* **1998**, *813*, 47–53. [CrossRef]
12. Radhika, S.; Nagaraju, V.; Kumar, B.N.; Kantam, M.L.; Reddy, B.R. Solid-liquid extraction of Gd(III) and separation possibilities of rare earths from phosphoric acid solutions using Tulsion CH-93 and Tulsion CH-90 resins. *J. Rare Earths* **2012**, *30*, 1270–1275. [CrossRef]
13. Kumar, B.N.; Radhika, S.; Reddy, B.R. Solid–liquid extraction of heavy rare-earths from phosphoric acid solutions using Tulsion CH-96 and T-PAR resins. *Chem. Eng. J.* **2010**, *160*, 138–144. [CrossRef]
14. Cheira, M.F. Characteristics of uranium recovery from phosphoric acid by an aminophosphonic resin and application to wet process phosphoric acid. *Eur. J. Chem.* **2015**, *6*, 48–56. [CrossRef]
15. Kabay, N.; Demircioglu, M.; Yayh, S.; Gunay, E.; Yuksel, M.; Saglam, M.; Streat, M. Recovery of Uranium from Phosphoric Acid Solutions Using Chelating Ion-Exchange Resins. *Ind. Eng. Chem. Res.* **1998**, *37*, 1983–1990. [CrossRef]
16. Flahaut, J. *Les éléments des Terres Rares*; Masson: Paris, France, 1969.
17. Reddy, B.R.; Kumar, B.N.; Radhika, S. Solid Liquid Extraction of Terbium from Phosphoric Acid Medium using Bifunctional Phosphinic Acid Resin Tulsion CH 96. *Solvent Extr. Ion Exch.* **2009**, *27*, 695–711. [CrossRef]
18. Bunus, F.; Dumitrescu, R. Simultaneous extraction of rare earth elements and uranium from phosphoric acid. *Hydrometallurgy* **1992**, *28*, 331–338. [CrossRef]

19. Li, D. A review on yttrium solvent extraction chemistry and separation process. *J. Rare Earths* **2017**, *35*, 107–119. [CrossRef]
20. Ochsentihn-Petropulu, M.; Lyberopulu, T.; Parissakis, G. Selective separation and determination of scandium from yttrium and lanthanides in red mud by a combined ion exchange/solvent extraction method. *Anal. Chim. Acta* **1995**, *315*, 231–237. [CrossRef]
21. Bao, S.; Hawker, W.; Vaughan, J. Scandium Loading on Chelating and Solvent Impregnated Resin from Sulfate Solution. *Solvent Extr. Ion Exch.* **2018**, *36*, 100–113. [CrossRef]
22. Alguacil, F.J. A kinetic study of cadmium(II) adsorption on Lewatit TP260 resin. *J. Chem. Res.* **2003**, *3*, 144–146. [CrossRef]
23. Rey, J.; Atak, S.; Dourdain, S.; Arrachart, G.; Berthon, L.; Pellet-Rostaing, S. Synergistic Extraction of Rare Earth Elements from Phosphoric Acid Medium using a Mixture of Surfactant AOT and DEHCNPB. *Solvent Extr. Ion Exch.* **2017**, *35*, 321–331. [CrossRef]

© 2018 by the authors. Licensee MDPI, Basel, Switzerland. This article is an open access article distributed under the terms and conditions of the Creative Commons Attribution (CC BY) license (http://creativecommons.org/licenses/by/4.0/).

Article

Effect of Aqueous Media on the Recovery of Scandium by Selective Precipitation

Bengi Yagmurlu [1,2,*], Carsten Dittrich [1] and Bernd Friedrich [2]

1. MEAB Chemie Technik GmBH, 52068 Aachen, Germany; carsten@meab-mx.com
2. RWTH Aachen, IME Institute of Process Metallurgy and Metal Recycling, 52056 Aachen, Germany; BFriedrich@metallurgie.rwth-aachen.de
* Correspondence: bengi@meab-mx.com; Tel.: +49-1575-4954583

Received: 9 April 2018; Accepted: 1 May 2018; Published: 3 May 2018

Abstract: This research presents a novel precipitation method for scandium (Sc) concentrate refining from bauxite residue leachates and the effect of aqueous media on this triple-stage successive precipitation process. The precipitation pattern and the precipitation behavior of the constituent elements was investigated using different precipitation agents in three major mineral acid media, namely, H_2SO_4, HNO_3, and HCl in a comparative manner. Experimental investigations showed behavioral similarities between HNO_3 and HCl media, while H_2SO_4 media was different from them because of the nature of the formed complexes. NH_4OH was found to be the best precipitation agent in every leaching media to remove Fe(III) with low Sc co-precipitation. To limit Sc loss from the system, Fe(III) removal was divided into two steps, leading to more than 90% of Fe(III) removal at the end of the process. Phosphate concentrates were produced in the final step of the precipitation process with dibasic phosphates which have a strong affinity towards Sc. Concentrates containing more than 50% of $ScPO_4$ were produced in each case from the solutions after Fe(III) removal, as described. A flow diagram of the selective precipitation process is proposed for these three mineral acid media with their characteristic parameters.

Keywords: bauxite residue; red mud; hydrometallurgy; recovery; scandium; precipitation

---

1. Introduction

The recent agreements and climate accords in reducing carbon emission and specified deadlines for automotive industries have placed light metals and alloys under the spotlight [1]. One of the reasons is the direct relation between a vehicle weight and its energy consumption. Scandium (Sc) is used as a tuning metal especially for aluminum alloys, which makes it one of the promising candidates for light-weight alloys [2]. It is, however, an extremely expensive element for widespread application in industrial usage at the moment [3]. As aluminum alloys with improved strength, thermal resistance, and weldability can be achieved with minor additions of Sc, improved oxygen-ion conductivity can also be attained in solid oxide fuel cells [4–6]. Hence, this metal was classified recently as a critical metal for the future, leading to a steep increase in its demand despite its price [7].

Unfortunately, Sc is widely dispersed in nature and generally has to be extracted from secondary raw materials or as a by-product of uranium, nickel laterite, or titanium pigment processing. Bauxite residue (i.e., red mud) is the by-product obtained through the Bayer Process, yielding approximately four billion tons, with a previously reported annual production of 160 million tons [8,9]. This alkaline waste can be considered as a valuable resource because of its metal content (Fe, Al, Ti, Sc, Rare earth elements (REEs), etc.). Therefore, the complete or partial valorization of bauxite residues (BR) has lately been of great interest [10–13].

Previously, complete or partial recovery of Sc from bauxite residues was reported to be achieved mainly by solvent extraction, ion exchange, or the combination of these two techniques, as a result of its low concentration in the leachates [14–20]. Zhang et al. recovered 91% of Sc from bauxite residue leachates by inorganic metal(IV)–phosphate ion exchangers, although Fe(III) was found to be an interfering ion in this process [20]. In another study, a newly developed supported ionic liquid phase (SILP) achieved almost complete Sc extraction, while showing a decreased efficiency in the presence of Fe(III) [14]. In all of these hydrometallurgical operations, the co-extraction of Fe, Al, and Ti became a problem, and intensive purification was required to produce a high-quality product.

In our previous work, a three-staged precipitation process was designed using a sulfuric acid media based on a selective Fe removal step by NH_4OH, since Fe is the most problematic element during Sc processing [21,22]. This was successively followed by selective Sc phosphate precipitation by $(NH_4)_2HPO_4$. As a result of this precipitation route, a Sc phosphate concentrate containing 65% Sc was synthesized from impure synthetic bauxite residue leach solutions. Nevertheless, the processing route must be tailored in relation to the geological presence of the bauxite residue, depending on different mineralogy and association of the phases. Hence, different mineral acids other than H_2SO_4, such as HNO_3 and HCl, were also tried for leaching bauxite residues.

In order to cover a wider range of bauxite residues as well as other waste-generating processes, such as Ti-pigment and Ni laterite production, the design of selective Sc precipitation route for bauxite residues has to be adapted to different mineral acid media. Thus, this paper investigates the effect of different aqueous media on the recovery of Sc by a selective precipitation method.

2. Materials and Methods

A bauxite residue sample was obtained from Aluminium of Greece, subjected to lithium borate fusion, and analyzed using inductively coupled plasma mass spectrometry, ICP-MS/AAS, as shown in Table 1.

Table 1. Chemical composition of the bauxite residue (LOI: Loss of ignition).

Major Compounds	wt. %	Minor Compounds	ppm
Fe	29.6	La	110
Al	8.6	Ce	380
Ca	8.3	Sc	120
Si	3.3	Nd	100
Ti	2.6	Y	80
Na	2.8		
Others	32.1		
LOI	12.7		
LOI: Loss on ignition			

Synthetic leach solutions were prepared considering pre-treatments before performing the selective precipitation. The expected leachate is the pregnant leach solutions (PLS) after the major part of Fe, Al, and Ti are recovered from the solution by both pyrometallurgical and hydrometallurgical methods. Hence, the synthetic solution mentioned in this study is predictive of the formation of the real PLS. The Sc and REEs concentrations were arranged to be higher than in the real PLS to clearly observe the precipitation behaviors of these elements. Additionally, the selective precipitation process was also tested directly on the real solutions, obtaining similar results as those observed in this study [21,23].

In addition to the major impurities present in the bauxite residue (e.g., Fe and Al), Nd and Y are also considered representative elements of light and heavy REEs within the synthetic PLS, since they have similar chemical properties as same sub-groups of REEs.

Synthetic solutions in chloride media were prepared by adding the required amount of reagent-grade $FeCl_3 \cdot 6H_2O$, $AlCl_3$, $Sc_2(SO_4)_3 \cdot 5H_2O$, $NdCl_3 \cdot 6H_2O$, and $YCl_3 \cdot 6H_2O$. In a similar manner, reagent-grade $Fe(NO_3)_3 \cdot 9H_2O$, $Al_2(SO_4)_3 \cdot 18H_2O$, $Sc_2(SO_4)_3 \cdot 5H_2O$, $NdCl_3 \cdot 6H_2O$, and $YCl_3 \cdot 6H_2O$ were used to synthesize the solution in nitrate media. Sulfate and chloride salts were first precipitated as hydroxides and washed before being converted into the necessary forms, to avoid unwanted sulfate or chloride ions which can affect the precipitation yields by promoting complex formations in the aqueous solution. All precipitation solutions were prepared from reagent-grade salts. The concentrations of the precipitation agents were 12.5 wt. % for $CaCO_3$ (limestone) slurry, 1 mol/L for NaOH, NH_4OH, and KOH, and 1 mol/L for K_2HPO_4, $(NH_4)_2HPO_4$, and Na_2HPO_4.

In order to have comparative precipitation results, a composition similar to the one used in the sulfate media was chosen. The pH of both systems was set between 1.2 and 1.4 with the addition of the necessary amounts of HCl or HNO_3. The composition of the synthetic solutions is presented in Table 2.

Table 2. Composition of the synthetic solutions in HNO_3 and HCl media.

	Composition in HNO_3 Media (mg/L)	Composition in HCl Media (mg/L)
Al	340	348
Fe(III)	312	290
Sc	86	78
Y	108	96
Nd	101	105

The precipitation agents mentioned were carefully added using a precision burette into 50 mL of the synthetic PLS while monitoring pH and temperature. All experiments presented in this study were done at room temperature. For hydroxide precipitation, the agents were added until the target pH was attained under mild agitation, to reach homogeneity in the solution and to prevent local pH differences.

Precipitation solutions containing dibasic phosphates were prepared as 1 mol/L and added into the leach solutions starting with a stoichiometric amount, considering only scandium precipitation. In each step, the amount added was doubled until reaching 20 times of the stoichiometric amount.

The resulted suspension for each case was then stabilized and homogenized at a given pH and temperature for 2 h and subsequently filtered through fine filter paper via suction filtration. The separated solid residue was washed with distilled water and dried at 110 °C for 24 h. Both filtered solutions and solid residues were assayed.

The concentrations of the constituent ions of iron (Fe), aluminum (Al), scandium (Sc), yttrium (Y), and neodymium (Nd) were determined by microwave plasma optical emission spectroscopy (Agilent MP-AES 4100, Mulgrave, VIC, Australia). Each sample was prepared by adding 100 µL of cesium ionization buffer and 500 µL of ultrapure concentrated HNO_3 to 10 mL of solution. Quantitative analyses were performed at 371.993 nm, 396.152 nm, 361.383 nm, 371.029 nm, and 430.358 nm, corresponding to the spectral emission lines for Fe, Al, Sc, Y, and Nd respectively.

The pH measurements were performed using a WTW ProfiLine pH 197 series pH-meter with a Sentix 81 precision electrode. The pH meter was calibrated with standard technical buffering solutions at pH 2.00, 4.01, and 7.00 to achieve maximum sensitivity in pH measurements.

3. Results

We previously reported that more than 90% of Fe can be removed with negligible amount of Sc loss from sulfuric acid-based solutions by simply adding NH_4OH in a dual-staged precipitation process [22]. Furthermore, with a successive precipitation route which combines both hydroxide precipitation and phosphate precipitation, a Sc phosphate concentrate, which is easier to process, can be synthesized from sulfate-based aqueous solutions. Since Sc-containing liquors can also exist in chloride or nitrate media, the effect of those aqueous media on precipitation should be investigated. By this way, the successive precipitation process can be adapted to other major mineral acid media.

3.1. Precipitation in HNO₃ Media

3.1.1. Fe Removal Step

In all recovery and purification operations regarding scandium, iron was reported to be one of the most problematic elements. Thus, to propose an easier route to process scandium, iron content should be minimized in the solution beforehand. We determined that the easiest route to this purpose was the hydroxide precipitation. Limestone slurry, sodium hydroxide, potassium hydroxide, and ammonium hydroxide were chosen as precipitation solutions, and a wide range of pH values was examined with the addition of these hydroxide donors.

The precipitation trends of the constituent elements as hydroxides are shown in Figure 1. The precipitation of Fe(III) with the addition of limestone slurry took place in the pH range between 2.0 and 3.5. As it can be seen from Figure 1a, when Fe(III) precipitation was triggered, the co-precipitation of the other components was around 25–30%. In the case of NaOH and KOH addition to the system (Figure 1b,c respectively), a more distinctive Fe(III) precipitation was observed. Yet, the co-precipitation levels were relatively high, between 10% and 20%, when more than 90% of Fe(III) was removed. The addition of NH₄OH produced similar results as those described in the sulfate system. A distinctive Fe(III) precipitation resulting in low co-precipitation levels of the other components was observed (Figure 1d). When more than 95% of Fe was removed from the solution, less than 5% of Sc and REEs precipitated. In all cases, the precipitation order was Fe(III) > Sc > Al > Y ≈ Nd.

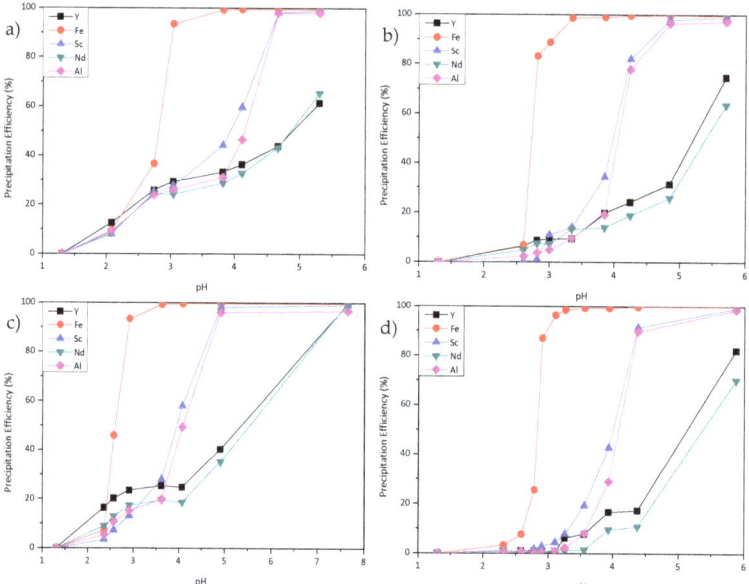

Figure 1. Precipitation behavior of Fe(III), Al, Sc, Nd, and Y in HNO₃ media at different pH values with the addition of (**a**) limestone; (**b**) NaOH; (**c**) KOH; (**d**) NH₄OH.

In this removal step, it is desirable to remove Fe from the system with minimum co-precipitation of the recoverable elements; therefore, selectivity was an essential parameter. The selectivity of the precipitation is calculated by Equation (1) given below:

$$D_{A,B} = \frac{C_{prec}}{C_{aq}} \text{ and } S_{A/B} = \frac{D_A}{D_B} \tag{1}$$

where D_A or D_B is the distribution coefficient of the mentioned element, C_{prec} is the concentration of the element in the precipitate, C_{aq} is the concentration of the element in the aqueous solution after precipitation, $S_{A/B}$ is the selectivity coefficient which indicates the selectivity of A over B.

The precipitation yields of the elements and the selectivity of Fe over Sc can be found in Table 3, when Fe precipitation reach 70% and 90%, respectively. Table 3a shows that the best candidates to remove Fe in HNO_3 media are NaOH and NH_4OH, since the corresponding $S_{Fe/Sc}$ values are far superior to those of the other hydroxide donors. Nevertheless, the co-precipitation levels of all elements abruptly increased in the case of NaOH upon further addition to remove 90% of Fe(III). Thus, NH_4OH showed an astonishing performance considering both the co-precipitation levels of the other elements and the $S_{Fe/Sc}$. Once that more than 95% of Fe present in the system was removed, around 4% of Sc and 1% of the other elements were precipitated with a remarkable $S_{Fe/Sc}$ of 585. The most logical explanation of this low co-precipitation levels is the occurrence of an hexamine scandium complex upon addition of NH_4OH [24,25]. Hence, Fe precipitation was triggered, whereas Sc remained in a complex form, which prevented the co-precipitation of Sc between these pH ranges.

The main difference of the nitrate-based aqueous solution compared to the sulfate-based one regards the pH ranges of the precipitation. Although it was found that the precipitation of Fe was triggered at pH values between 2.5 to 4.0 in sulfate media, a similar precipitation level was observed at a lower pH, between 2.0 to 3.0. It is known that sulfate ion form inner-sphere complexes, while nitrate complexes can be classified as forming outer-sphere complexes [26,27]. Consequently, more OH^- ions have to be released to disrupt inner-sphere complexes, since they have lower Gibbs Energy (ΔG) in that state.

Table 3. (a) Critical pH values for 70% Fe(III) removal from the system by hydroxide precipitation and precipitation % of the constituent ions with selectivity of Fe over Sc; (b) Critical pH values to obtain above 90% Fe(III) removal from the system by hydroxide precipitation and precipitation % of the constituent ions with selectivity of Fe over Sc.

			(a)				
Precipitation Agent	pH	Fe (%)	Sc (%)	Al (%)	Y (%)	Nd (%)	$S_{Fe/Sc}$
Limestone	2.95	76.6	26.8	25.1	28.1	24.2	9
NaOH	2.78	72.3	1.1	3.3	8.1	6.8	235
KOH	2.74	69.7	10.4	13.1	21.9	15.2	20
NH_4OH (aq)	2.87	71.5	2.4	0.5	0.5	0.5	102
			(b)				
Precipitation Agent	pH	Fe (%)	Sc (%)	Al (%)	Y (%)	Nd (%)	$S_{Fe/Sc}$
Limestone	3.05	93.6	27.6	25.9	29.4	24.2	38
NaOH	3.00	88.9	11.0	5.1	9.3	7.4	65
KOH	2.91	93.4	13.3	15.3	23.5	17.4	92
NH_4OH (aq)	3.10	96.5	4.5	0.9	0.6	1.3	585

3.1.2. Phosphate Precipitation

In previous studies, it was shown that there is a strong affinity of Sc and REEs for PO_4^{3-} ion [22,28,29]. In leach solutions with sulfate media, addition of dibasic phosphate resulted in a selective precipitation of both Fe(III) and Sc from the system. In light of this, dibasic phosphate solutions were tested with the intention of recovering Sc from the PLS. Addition of three different dibasic phosphate solutions and the resulting precipitation patterns of the constituent elements are shown in Figure 2. Precipitation solutions were added starting from the stoichiometric amount considering only Sc precipitation, and the amount introduced was increased in each step until reaching an amount 20 times greater than the stoichiometric value.

In all cases, similar precipitation trends were observed. Upon addition of the precipitant solution, immediate precipitation of Fe and Sc was initiated. More than 90% of Sc and Fe precipitation occurred at a pH value around 2.5. Since dibasic phosphates release OH^- into the system, an uncontrolled rise of pH would lead to the precipitation of the constituent elements as hydroxides, which will deteriorate the selectivity towards Sc recovery. It is important to note that, while high levels of co-precipitation of the other elements were detected with Na$_2$HPO$_4$ addition, corresponding approximately to 40–50%, it was found that they were limited between 20% and 25% when K$_2$HPO$_4$ or (NH$_4$)$_2$HPO$_4$ were introduced into the solution as phosphate donors. In all cases, the additions yielded to the precipitation order Fe(III) = Sc > Al > Y = Nd. The precipitation efficiencies upon the addition of dibasic phosphate precipitation solutions can be found in Table 4.

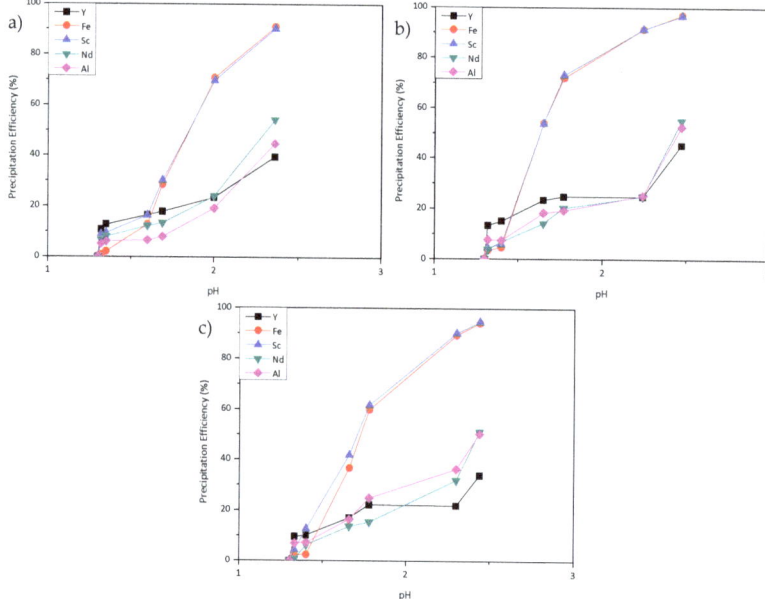

Figure 2. Precipitation of the elements in HNO$_3$ media with the addition of (**a**) Na$_2$HPO$_4$; (**b**) K$_2$HPO$_4$ and (**c**) (NH$_4$)$_2$HPO$_4$.

Table 4. Critical pH values for above 90% Sc recovery from the solution system by dibasic phosphate precipitation and precipitation efficiencies of the constituent ions.

Precipitation Agent	pH	Fe(III) (%)	Sc (%)	Al (%)	Y (%)	Nd (%)
Na$_2$HPO$_4$	2.36	90.8	90.2	44.8	39.6	54.3
K$_2$HPO$_4$	2.24	91.2	91.3	25.1	24.8	24.9
(NH$_4$)$_2$HPO$_4$	2.30	89.3	90.2	36.3	22.0	31.2

3.2. Precipitation in HCl Media

3.2.1. Fe Removal Step

The same precipitation solutions for hydroxide precipitation were further tested in HCl media, and the results are summarized in Figure 3. Similar precipitation routes and behaviors were observed as in the case of HNO$_3$ media. While the addition of limestone slurry resulted in the co-precipitation of all other elements with Fe(III), more distinctive cases were achieved when NaOH, KOH, and NH$_4$OH were used. As in the previous case, with the addition of NH$_4$OH, the lowest co-precipitation was achieved, yet Sc precipitation was observed to be the greatest among the other elements when the Fe content in the solution was minimized. In all cases, the precipitation order was Fe(III) > Sc > Al > Y = Nd.

The precipitation percentages of the elements in the solution can be found in Table 5, when Fe(III) was removed at 70% and 90%, respectively. The major difference in the precipitation behavior between HNO$_3$ media and HCl media regarded the co-precipitation levels of Sc. It was previously discussed that both in nitrate and in sulfate media, Sc remained in the solution, while Fe(III) was almost completely taken out from the solution. With the addition of NH$_4$OH, 70% of Fe(III) as well as 9% of Sc were precipitated at pH values around 2.8, while Sc precipitation remained at 13% once 93% of Fe(III) was removed from the solution at a pH value around 3.0. The interaction between Fe^{3+} and Cl$^-$ ions was the main reason why the precipitation level was limited, and the co-precipitation level of Sc increased when Fe started to precipitate. This interaction shifted the precipitation range of Fe(III) slightly, prompting Sc precipitation in the same range.

Still, the precipitation procedure to synthesize a scandium concentrate is applicable even in this case. Although the selectivity of Fe over Sc was found to be 24 when the precipitation of Fe reached 70%, as the precipitation progressed, a value of 100 was reached with the addition of NH$_4$OH.

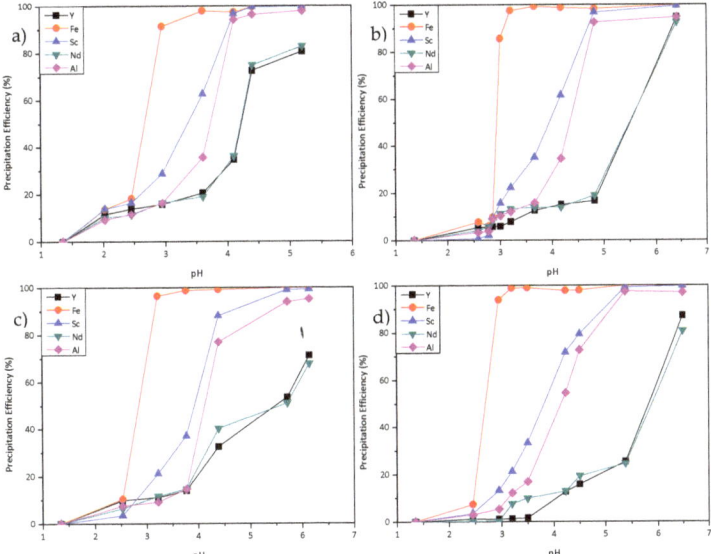

Figure 3. Precipitation behavior of Fe(III), Al, Sc, Nd, and Y in HCl media with the addition of (a) CaCO$_3$; (b) NaOH; (c) KOH; (d) NH$_4$OH.

Table 5. (a) Critical pH values for 70% Fe(III) removal from the system by hydroxide precipitation and precipitation % of the constituent ions with selectivity of Fe over Sc; (b) Critical pH values for above 90% Fe(III) removal from the system by hydroxide precipitation and precipitation % of the constituent ions with selectivity of Fe over Sc.

(a)							
Precipitation Agent	pH	Fe (%)	Sc (%)	Al (%)	Y (%)	Nd (%)	$S_{Fe/Sc}$
Limestone	2.81	70.7	25.2	15.3	15.1	15.6	7
NaOH	2.97	70.3	14.4	9.9	5.8	10.6	14
KOH	3.05	75.1	16.7	8.2	10.6	8.3	15
NH$_4$OH	2.82	71.3	9.6	4.4	1.0	0.2	24
(b)							
Precipitation Agent	pH	Fe (%)	Sc (%)	Al (%)	Y (%)	Nd (%)	$S_{Fe/Sc}$
Limestone	2.95	91.2	28.6	25.9	29.4	24.2	25
NaOH	3.10	91.4	18.9	11.1	6.9	12.3	46
KOH	3.22	96.3	21.0	9.1	11.1	11.5	98
NH$_4$OH	2.95	93.8	13.2	5.2	1.1	0.3	100

3.2.2. Phosphate Precipitation

Figure 4 shows the precipitation behavior with the addition of different phosphate donors into a synthetic solution with HCl media. The selective precipitations of scandium and iron were triggered immediately with the addition of a dibasic phosphate solution, and 60% precipitation efficiency was attained, with below 10% co-precipitation of other elements at pH values around 1.5 in all cases. After that point, further addition of phosphate ions resulted in the growth of both targeted and non-targeted elements. The best selectivity with 90% of Sc precipitation efficiency was reached at pH 2.2 with the addition of K_2HPO_4, with co-precipitation yields of 27%, 26%, and 43% for Al, Y, and Nd, respectively. $(NH_4)_2HPO_4$ showed a similar performance with slightly increased co-precipitation levels corresponding to 36% for Al, 23% for Y, and 42% for Nd when Sc precipitation hit 90%. Critical pH values for above 90% Sc recovery from the solution system by dibasic phosphate precipitation and the precipitation efficiencies of the constituent ions can be found in Table 6.

Since introducing new ions into the system is unwanted throughout the precipitation process, $(NH_4)_2HPO_4$ was selected as the best precipitation agent for selective precipitation of Sc from bauxite residue leachates.

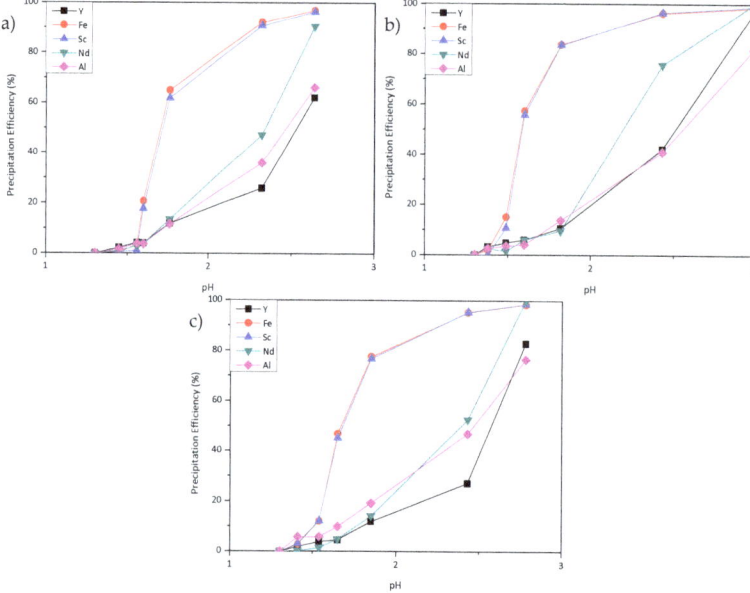

Figure 4. Precipitation of the elements in HCl media with the addition of (**a**) Na_2HPO_4; (**b**) K_2HPO_4 and (**c**) $(NH_4)_2HPO_4$.

Table 6. Critical pH values for above 90% Sc recovery from the solution system by dibasic phosphate precipitation and precipitation efficiencies of the constituent ions.

Precipitation Agent	pH	Fe(III) (%)	Sc (%)	Al (%)	Y (%)	Nd (%)
Na_2HPO_4	2.32	92.1	90.9	39.2	26.0	47.3
K_2HPO_4	2.19	90.1	90.2	27.5	26.3	42.7
$(NH_4)_2HPO_4$	2.27	90.4	90.3	36.3	22.9	42.1

3.3. Successive Precipitation in All Media

Taking all the findings into consideration, triple-staged precipitation processes for the synthesis of a scandium concentrate from bauxite residue solutions can be proposed using the described mineral acid systems. The proposed precipitation processes are summarized in Figure 5, and the variation in concentration with the successive addition of the precipitation agents in H_2SO_4, HNO_3, and HCl media can be found in Figure 5a–c, respectively. Three precipitation regions labelled as regions 1, 2, and 3 are shown in all graphs. Region 1 is the removal of Fe from a solution with minimum Sc loss by the addition of NH_4OH until a specified pH is reached for each media. The residue obtained from this step, which was enriched in Fe, was filtered and removed from the solution. The critical pH ranges for the first Fe removal step were established as 3.3–3.4 for H_2SO_4 and 2.8–2.9 for HNO_3 and HCl. Region 2 denotes the second Fe removal step by further addition of NH_4OH that promotes higher Fe precipitation with a low amount of Sc co-precipitation. The precipitate was then filtered and could be recycled into the initial feed, which minimized Sc losses during the second Fe removal step and provided a seeding agent for better Fe separation. The pH ranges for this step were between 3.6–3.7 for H_2SO_4, 3.1–3.2 for HNO_3, and 3.0–3.1 for HCl media.

After removing more than 95% of Fe using NH_4OH with low Sc loss, a phosphate precipitation step was applied in the third part of this successive precipitation process. It was observed that the phosphate precipitation with dibasic phosphates showed similar performances regardless of the system. The pH of all systems was adjusted to 2.0 in advance, with the purpose to avoid unwanted hydroxide precipitation which can be triggered at pH levels above 3.0. Dibasic phosphate solutions were added into all systems until reaching a pH range between 2.5 and 2.6, where the best selectivity for Sc was reached. Since Fe was removed in previous steps, both the amount of precipitant solution and the co-precipitation levels of the constituent ions were decreased. For all media, more than 95% of Sc and Fe was recovered, while approximately 15% of Al and 10% of Y and Nd were co-precipitated during the process. The compositions of the resulted concentrate in the form of mixed phosphates can be found in Table 7.

Table 7. Compositions of each concentrate obtained after their successive selective precipitation as phosphates.

	H_2SO_4	HNO_3	HCl
$ScPO_4$ (%)	66	56	51
$FePO_4$ (%)	13	12	13
$AlPO_4$ (%)	19	23	26
YPO_4 (%)	1	3	5
$NdPO_4$ (%)	1	6	5

In all cases, Sc concentrates containing more than 50% $ScPO_4$ were synthesized. Since the precipitation patterns of the ions were more discrete in the sulfuric acid media, the best results were obtained in this media, with 66% $ScPO_4$. The major impurity element was found to be Al in the concentrate, which can be easily processed and removed with basic purification operations.

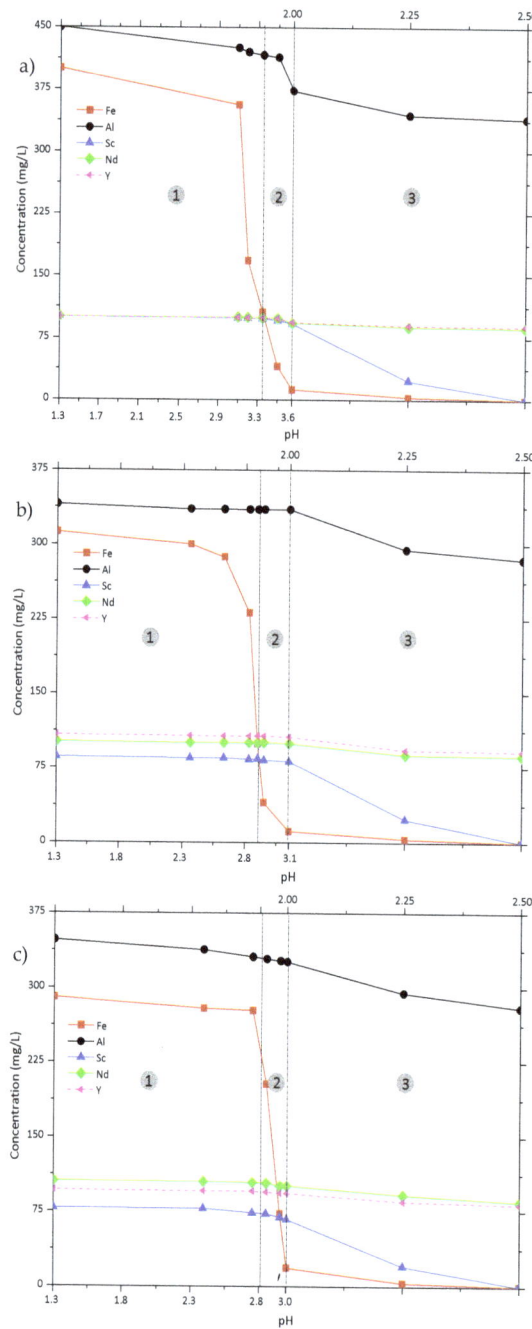

Figure 5. Triple-staged successive precipitation with NH$_4$OH and (NH$_4$)$_2$HPO$_4$ from (**a**) H$_2$SO$_4$; (**b**) HNO$_3$ and (**c**) HCl media.

4. Assessment and Conclusions

A triple-step precipitation route is proposed to refine a scandium concentrate from synthetic bauxite residue leachates, and the effect in aqueous media was investigated. In all media, NH$_4$OH showed better selectivity and performance than the other hydroxides, similar to the results obtained in H$_2$SO$_4$ media. While HNO$_3$ and HCl media showed quite similar patterns during precipitation, they differed with respect to the H$_2$SO$_4$ media, especially in the pH ranges of precipitation, because of the nature of the formed complexes. Although Sc loss during the Fe removal step in HCl media was observed to be relatively higher compared to the other cases as a result of the interaction between Fe^{3+} and Cl$^-$ ions, still the concentrate contained more than 50% of Sc. The proposed selective precipitation route to produce ScPO$_4$ concentrate can be seen in Figure 6.

A scandium concentrate containing more than 50% of ScPO$_4$ was successfully obtained with this selective precipitation route from synthetic solutions. Depending on the impurity level, the aqueous media, and the initial composition of the feed solution, the amount of ScPO$_4$ in the synthesized concentrate could vary between 15% and 65%. Since the concentrate contained more scandium than the other elements, the processing and the purification of this concentrate will be much easier to perform. The removal of the most problematic element for scandium precipitation, which is iron, is the main obstacle for advanced processes to obtain a high-purity scandium product. Furthermore, the remaining solution after precipitation can be treated for further REEs recovery through other means.

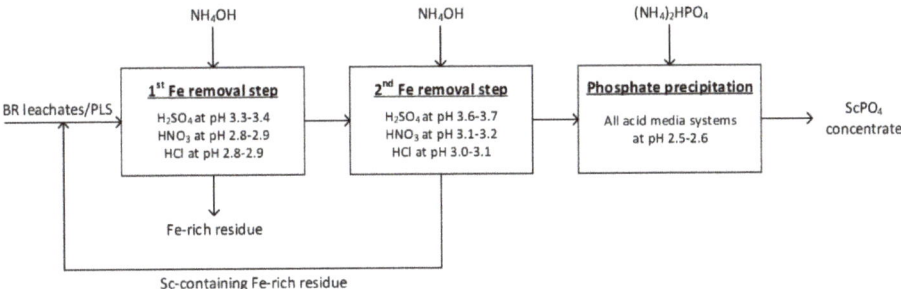

Figure 6. Proposed selective precipitation route to synthesize a ScPO$_4$ concentrate using three mineral acid media.

Author Contributions: B.Y., C.D., and B.F. conceived and designed the experiments; B.Y. performed the experiments; B.Y. analyzed the data; B.Y. wrote the paper; B.Y., C.D., and B.F. reviewed the paper.

Funding: This research was funded by European Community's Horizon 2020 Programme under grant agreement number 636876.

Acknowledgments: The research leading to these results has received funding from the European Community's Horizon 2020 Programme ([H2020/2014–2019]) under Grant Agreement no. 636876 (MSCA-ETN REDMUD). This publication reflects only the author's view, exempting the Community from any liability. Project website: http://www.etn.redmud.org. The authors thank Wenzhong Zhang and Dzenita Avdibegovic for their supports in ICP Measurements.

Conflicts of Interest: The authors declare no conflict of interest.

References

1. UNFCCC. Adoption of the Paris Agreement. Report No. FCCC/CP/2015/L.9/Rev.1. Available online: http://unfccc.int/resource/docs/2015/cop21/eng/l09r01.pdf (accessed on 15 January 2018).
2. Røyset, J.; Ryum, N. Scandium in aluminium alloys. *Int. Mater. Rev.* **2005**, *50*, 19–44. [CrossRef]
3. Gambogi, J. *USGS Minerals Information: Scandium*; U.S. Geological Survey: Reston, VA, USA; pp. 146–147.
4. Lathabai, S.; Lloyd, P. The effect of scandium on the microstructure, mechanical properties and weldability of a cast Al–Mg alloy. *Acta Mater.* **2002**, *50*, 4275–4292. [CrossRef]

5. Marquis, E.; Seidman, D. Nanoscale structural evolution of Al$_3$Sc precipitates in Al-Sc alloys. *Acta Mater.* **2001**, *49*, 1909–1919. [CrossRef]
6. Yamamoto, O. Solid oxide fuel cells: Fundamental aspects and prospects. *Electrochim. Acta* **2000**, *45*, 2423–2435. [CrossRef]
7. European Commission. *Study on the Review of the List of Critical Raw Materials: Executive Summary*; Directorate-General for Internal Market, Industry, Entrepreneurship and SMEs: Brussels, Belgium, 2017; pp. 1–93.
8. Power, G.; Gräfe, M.; Klauber, C. Bauxite residue issues: I. Current management, disposal and storage practices. *Hydrometallurgy* **2011**, *108*, 33–45. [CrossRef]
9. Evans, K. The history, challenges, and new developments in the management and use of bauxite residue. *J. Sustain. Metall.* **2016**, *2*, 316–331. [CrossRef]
10. Li, G.; Liu, M.; Rao, M.; Jiang, T.; Zhuang, J.; Zhang, Y. Stepwise extraction of valuable components from red mud based on reductive roasting with sodium salts. *J. Hazard. Mater.* **2014**, *280*, 774–780. [CrossRef] [PubMed]
11. Liu, Y.; Naidu, R. Hidden values in bauxite residue (red mud): Recovery of metals. *Waste Manag.* **2014**, *34*, 2662–2673. [CrossRef] [PubMed]
12. Paramguru, R.; Rath, P.; Misra, V. Trends in red mud utilization—A review. *Miner. Process. Extr. Metall. Rev.* **2004**, *26*, 1–29. [CrossRef]
13. Alkan, G.; Yagmurlu, B.; Cakmakoglu, S.; Hertel, T.; Kaya, Ş.; Gronen, L.; Stopic, S.; Friedrich, B. Novel approach for enhanced scandium and titanium leaching efficiency from bauxite residue with suppressed silica gel formation. *Sci. Rep.* **2018**, *8*, 5676. [CrossRef] [PubMed]
14. Avdibegović, D.; Regadío, M.; Binnemans, K. Recovery of scandium(III) from diluted aqueous solutions by a supported ionic liquid phase (silp). *RSC Adv.* **2017**, *7*, 49664–49674. [CrossRef]
15. Narayanan, R.P.; Kazantzis, N.K.; Emmert, M.H. Selective process steps for the recovery of scandium from jamaican bauxite residue (red mud). *ACS Sustain. Chem. Eng.* **2018**, *6*, 1478–1488. [CrossRef]
16. Ochsenkühn-Petropulu, M.; Lyberopulu, T.; Parissakis, G. Selective separation and determination of scandium from yttrium and lanthanides in red mud by a combined ion exchange/solvent extraction method. *Anal. Chim. Acta* **1995**, *315*, 231–237. [CrossRef]
17. Onghena, B.; Borra, C.R.; Van Gerven, T.; Binnemans, K. Recovery of scandium from sulfation-roasted leachates of bauxite residue by solvent extraction with the ionic liquid betainium bis (trifluoromethylsulfonyl) imide. *Sep. Purif. Technol.* **2017**, *176*, 208–219. [CrossRef]
18. Wang, W.; Cheng, C.Y. Separation and purification of scandium by solvent extraction and related technologies: A review. *J. Chem. Technol. Biotechnol.* **2011**, *86*, 1237–1246. [CrossRef]
19. Wang, W.; Pranolo, Y.; Cheng, C.Y. Recovery of scandium from synthetic red mud leach solutions by solvent extraction with D2HPA. *Sep. Purif. Technol.* **2013**, *108*, 96–102. [CrossRef]
20. Zhang, W.; Koivula, R.; Wiikinkoski, E.; Xu, J.; Hietala, S.; Lehto, J.; Harjula, R. Efficient and selective recovery of trace scandium by inorganic titanium phosphate ion-exchangers from leachates of waste bauxite residue. *ACS Sustain. Chem. Eng.* **2017**, *5*, 3103–3114. [CrossRef]
21. Yagmurlu, B.; Alkan, G.; Xakalashe, B.; Friedrich, B.; Stopic, S.; Dittrich, C. Combined saf smelting and hydrometallurgical treatment of bauxite residue for enhanced valuable metal recovery. In Proceedings of the 35th International Conference and Exhibition of ICSOBA, Hamburg, Germany, 2–5 October 2017.
22. Yagmurlu, B.; Dittrich, C.; Friedrich, B. Precipitation trends of scandium in synthetic red mud solutions with different precipitation agents. *J. Sustain. Metall.* **2017**, *3*, 90–98. [CrossRef]
23. Alkan, G.; Xakalashe, B.; Yagmurlu, B.; Kaussen, F.; Friedrich, B. Conditioning of red mud for subsequent titanium and scandium recovery–a conceptual design study. *World Metall. ERZMETALL* **2017**, *70*, 5–12.
24. Horovitz, C.T. *Scandium its Occurrence, Chemistry Physics, Metallurgy, Biology and Technology*; Elsevier: New York, NY, USA, 2012.
25. Vickery, R.C. *The Chemistry of Yttrium and Scandium*; Oxford: Oxford, UK, 1960; Volume 2.
26. Ahrland, S. How to distinguish between inner and outer sphere complexes in aqueous solution. Thermodynamic and other criteria. *Coord. Chem. Rev.* **1972**, *8*, 21–29. [CrossRef]
27. Spiro, T.G.; Revesz, A.; Lee, J. Volume changes in ion association reactions. Inner-and outer-sphere complexes. *J. Am. Chem. Soc.* **1968**, *90*, 4000–4006. [CrossRef]

28. Lucas, S.; Champion, E.; Bregiroux, D.; Bernache-Assollant, D.; Audubert, F. Rare earth phosphate powders REPO$_4$·NH$_2$O (Re = La, Ce or Y)—Part i. Synthesis and characterization. *J. Solid State Chem.* **2004**, *177*, 1302–1311. [CrossRef]
29. Beltrami, D.; Deblonde, G.J.-P.; Bélair, S.; Weigel, V. Recovery of yttrium and lanthanides from sulfate solutions with high concentration of iron and low rare earth content. *Hydrometallurgy* **2015**, *157*, 356–362. [CrossRef]

© 2018 by the authors. Licensee MDPI, Basel, Switzerland. This article is an open access article distributed under the terms and conditions of the Creative Commons Attribution (CC BY) license (http://creativecommons.org/licenses/by/4.0/).

Article

Europium, Yttrium, and Indium Recovery from Electronic Wastes

Ernesto de la Torre *, Estefanía Vargas, César Ron and Sebastián Gámez

Department of Extractive Metallurgy, Escuela Politécnica Nacional, Ladrón de Guevara E11-253, Quito 170517, Ecuador; maria.vargas01@epn.edu.ec (E.V.); cesar.ron.1991@gmail.com (C.R.); sebastian.gamez@epn.edu.ec (S.G.)
* Correspondence: ernesto.delatorre@epn.edu.ec; Tel.: +593-(9)9947-1051

Received: 15 September 2018; Accepted: 27 September 2018; Published: 29 September 2018

Abstract: Waste electrical and electronic equipment (WEEE) has increased in recent decades due to the continuous advancement of technology in the modern world. These residues have various metals that are found in concentrations that make their recovery profitable. A group of metals of interest are the rare earths such as europium and yttrium, as well as semiconductors such as indium. Yttrium was recovered from cathode ray tubes that were manually dismantled. The resulted powder was leached with HNO_3, and then the solution was submitted to solvent extraction with di-(2-ethylhexyl) phosphoric acid (DEHPA) using *n*-heptane as a diluent. For re-extraction, HNO_3 was used again, and yttrium was precipitated by adding four times the stoichiometric amount of oxalic acid, reaching 68% yttrium purity. Indium was recovered from the liquid crystal display (LCD) screens for which the pulverized material was leached with H_2SO_4. Then, the indium sulfate was subjected to solvent extraction using DEHPA as an extractant, and diesel as a diluent. The re-extraction was carried out again with H_2SO_4, and the obtained acid solution was evaporated until the indium precipitated, reaching a recovery of 95%. The investigations that were carried out show that it is feasible to recover these metals in the form of oxides or phosphates with high commercial value.

Keywords: WEEE; yttrium; indium; hydrometallurgy

---

1. Introduction

During the last decades, there has been a vertiginous development of the technological industry, which has caused a significant increase in electronic waste. Electric and electronic wastes, which are also called waste electrical or electronic equipment (WEEE), is a general term that covers any device that has ceased to be useful [1]. The electronic waste that is generated due to the continuous advance of the technology is stored in deposits as scrap. However, many of these WEEE have high value-added metals in their structure in concentrations that are much higher than those within minerals [2].

Globally, WEEE is the waste that has shown the most growth in recent decades. WEEE constitutes approximately 8% of total urban waste [3]. It is estimated that 14 kg of WEEE per year are generated per person, and this tendency grows exponentially each year [4]. Despite this trend, there are no regulations or provisions for the proper disposal of WEEE in undeveloped countries due to economic constraints. In addition, it must be taken into account that the number of electronic devices and the recycling rate is lower than in developed countries. In the case of Europe, it is known that the amount of WEEE is approximately 7.5 million tons, and the annual growth of this trend ranges between 3–5% [5].

Another aspect to take into account is the negative environmental impact caused by WEEE if they are not properly disposed. When electronic devices end their useful life, they are destined for landfills or incinerators. However, it must be taken into account that WEEE has toxic metals such as lead, cadmium, or mercury, which are harmful to the environment and the health of living beings [6].

It should be noted that one of the main attractions of electronic waste is the high concentration of precious metals, base metals, and rare earths, which makes them potential sources of these elements. For instance, it is common to find 41% and 22% of iron and copper respectively within WEEE, since they are the major elements in electronic wastes. Precious metals and rare earths are in ppm concentrations instead [7,8]. Much electronic waste has precious metals or rare earths in concentrations that make it possible to extract them. One of them is cathode ray tubes (CRTs), which present a powder composed of zinc sulfide and yttrium oxysulfide activated with europium. The CRTs are found on computer screens and televisions that are covered by layers of barium oxide and lead that protect users from the exposure of X-rays generated within the CRTs. It should be noted that the glasses that are part of the CRTs have up to 28% lead, which becomes a serious environmental problem when the CRTs are leached in the landfills where they are stored [6].

However, due to the presence of yttrium and europium in its structure, the CRTs are highly appreciated, since rare earths have a high commercial value [9].

The elements of rare earths are a group composed of 17 chemical elements, which includes the 15 metals corresponding to the series of lanthanides in addition to yttrium and scandium. All of these elements present similarities in their chemical, electrical, magnetic, and optical properties [10]. This similarity is mainly due to the electronic configuration of the rare earth elements. In the atoms of these elements, f orbitals are partially filled, and the electrons are added to sublayer 4f, which is surrounded by 6s sublayers. Since the last layer is completely filled, in the case of all of the elements, the electrons of sublayer 4f are so well protected that the chemical properties remain practically unchanged. In the case of lanthanum, there are no electrons in the f orbitals, but the 5d sublayer is the one that is protected by the filled external sublayers [11,12]. Due to their exceptional physical, optical, electrical, and magnetic properties, these elements are used in a wide range of applications, from metal alloys, flat screens, hard drives, and oil-refining catalysts to medical equipment. There are important uses in the defense industry, such as in fighter aircraft engines, guided missile systems, missile defense units, and satellite communication systems. According to their use, these elements are used as mixtures, in addition to other chemical compounds, or in metallic and alloy forms [13].

The main mineral sources of rare earths are bastnasite [(Ce,La)(CO$_3$)F], monazite [(Ce,La)PO$_4$)], and xenotime (YPO$_4$). These minerals are submitted to a comminution process followed by a gravimetric concentration process [14,15]. The processed minerals that are enriched in rare earths are leached with acids or alkalis. Generally, hydrochloric (HCl) or sulfuric acid (H$_2$SO$_4$) is employed at this stage, and according the leaching agent concentration, pulp density, and temperature, rare earths may have selective solubilities that may make their separation from other metals possible [16,17].

Rare earths that are dissolved after leaching are recovered by several hydrometallurgical techniques such as: solvent extraction, ion exchange, and crystallization. The most used technique is solvent extraction, since extractants such as di-(2-ethylhexyl) phosphoric acid (DEHPA) enable obtaining products of high purity [18]. Re-extraction is accomplished mainly with the employment of acids such as HCl. Several parameters must be optimized, such as the molar ratio between the organic and aqueous phase, extractants concentration, and the contact time between the two phases, among others. In the case of yttrium, it can be recovered from the aqueous phase by precipitation after the addition of oxalic acid or sodium oxalate. Its transformation to an oxide can be achieved by calcination at 750 °C or, it can be heated to 750 °C with the addition of hydrofluoric acid in order to obtain yttrium fluoride [19].

Another group of electronic waste that has several elements of commercial interest is liquid crystal displays (LCDs). These devices are used in the manufacture of televisions, laptops, cameras, and mobile phones, among others [20]. The LCD screens have an approximate lifespan of up to eight years, so once their use is over, they must be properly disposed due to the presence of several toxic elements in their structure, such as mercury. Approximately 72% of LCD screens have fluorescent cathode lamps that contain mercury in their structure. One of the elements that has great commercial value and is inside the LCD screens is indium. This element is a semiconductor that has several properties such as electrical conductivity, chemical stability, and high hardness. The indium that is found in the liquid crystal panel has great commercial appeal when it is sold as indium oxide [20,21].

Indium is commonly found in the veins of minerals that are associated with sulfides and have a high content of tin. Several places have been discovered where indium deposits are found, such as: Kidd Creek, Timmins, ON, Canada (0.027% w/w indium); Polaris, Nunavut, NT, Canada (0.010% w/w indium); Balmat, NY, USA (0.004% w/w indium); and Toyoha, Japan (6% w/w indium) [22]. There is also a massive sulfur deposit housed in the volcano of the Kidd Creek mine, with an indium content between 1–870 ppm, with an average of 106 ppm that can be exploitable. Tin and tungsten deposits house the highest concentrations of indium [23].

Indium is produced mainly from the waste obtained from the zinc refining and the recycling of chimney dust, slag, and gases generated during the zinc smelting, where the degree of indium is 0.027%. These residues are subjected to a process of leaching through the use of hydrochloric (HCl) or sulfuric acid (H_2SO_4). The solutions are concentrated by solvent extraction, and the indium is recovered by electrodeposition as 99% metal. The low grade of indium is then refined to a standard grade metal (99.99%) of higher purity [24]. The concentrate is processed by roasting to eliminate sulfur, and a leaching process is carried out where the iron is eliminated as a residue of jarosite. In order to recover the indium, the jarosite residue is dissolved using a hot sulfuric acid solution. In this method, the indium is extracted directly from the leaching solution by solvent extraction. This technique is not suitable for residues of zinc plants with low concentrations of indium [25].

The present research analyzes the recovery of europium, yttrium, and indium from WEEE through the hydrometallurgical route. The main objective of the work consists of recovering yttrium and indium as suitable salts for their commercialization. Several works during the last years have attempted to recover yttrium and indium from electric and electronic wastes. Yang et al. [26] tried to separate yttrium and indium through employing acid leaching and solvent extraction techniques. In this case, di-(2-ethylhexyl) phosphoric acid (DEHPA) was used as the extractant and kerosene was used as the diluent, achieving 99% of yttrium and indium recovery in 2 M HCl and 1 M HNO_3 solutions, respectively [26]. On the other hand, De Michelis et al. [27] and Innocenzi et al. [28] performed rare earths recovery from WEEE by acidic leaching and precipitation as the main techniques. De Michelis et al. [27] employed several acids to dissolve rare earths with a subsequent precipitation with oxalic acid. This technique allowed obtaining yttrium oxalate (99% of recovery). Innocenzi et al. [28] instead attempted yttrium recovery using NaOH solutions for yttrium precipitation from acidic solutions. In this research, 95% of yttrium recovery was achieved. These research studies demonstrated that rare earths and indium extractions from WEEE are possible nowadays.

Considering recent research studies in rare earths and indium extraction from WEEE, this work focused on the extraction of the mentioned elements from cathode ray tubes (CRTs) and liquid crystal displays (LCD) panels. Unlike other investigations in this area, three techniques were tested in order to obtain yttrium and indium as commercial salts; these hydrometallurgical techniques are: acidic leaching, solvent extraction, and precipitation. The novelty of this work lies in the employment of CRTs and LCD panels as rare earths and indium sources, respectively. In both studies, several acids were tested in order to achieve the greatest element dissolution. Afterwards, several extractants and different diluents were used in order to separate the desired elements from other impurities such as zinc, iron, and aluminum, among others. In contrast with other research studies, solvent extraction was implemented instead of attempting a selective precipitation of rare earths and indium, which may create products with several impurities. Finally, yttrium oxide and indium sulfate were obtained as the main products of this investigation. The results obtained show that the adequate disposal of WEEE for the recovery of metals with high added value is feasible.

2. Materials and Methods

2.1. Materials

The following substances were employed in the development of this research: Nitric acid (HNO_3), 68% Panreac; sulfuric acid (H_2SO_4), 98% Mallinckrodt; hydrochloric acid (HCl), 37% Panreac; hydrofluoric acid (HF), Merck; n-hexane, 99.9% JT Baker; n-heptane, 99.9% Merck; di-(2-ethylhexyl) phosphoric acid (DEHPA), 99.9% Merck; tributyl phosphate (TBP), 97%, Fluka AG; methyl trioctyl ammonium chloride (Aliquat 336), 90% Merck; oxalic acid dehydrate, 99% Merck; and diesel premium and sodium hydroxide (NaOH), technical grade.

2.2. Methodology for Europium and Yttrium Recovery

Powder from CRTs was extracted manually and collected with a vacuum cleaner. The powder was characterized with an X-ray diffractometer Bruker AXS D8 Advance model and a scanning electron microscopy (SEM)–electric dispersive scanning (EDS) Vega TESCAN. Leaching essays were performed with HNO_3, H_2SO_4, and HCl for yttrium and europium dissolution under magnetic stirring at different periods of time. Various concentrations of the leaching agent (100–350 g/L) were assessed as well as different pulp densities (10%, 20%, 30% of solids) in order to determine the optimal parameters. The amount of dissolved yttrium and europium was measured by atomic absorption spectroscopy with a Perkin Elmer AAnalyst 300 spectrometer. The error that was associated to the atomic absorption analysis was ±0.1%. The amount of undissolved yttrium and europium was determined by atomic absorption, for which the CRT powder was submitted to acid disintegration with HNO_3, HF, and HCl. Rare earths recoveries were obtained by making a mass balance as follows (see Appendix A):

$$\text{Rare earth recovery} = \frac{m_{\text{rare earth in solution}} \,(mg)}{m_{\text{rare earth in solution}} \,(mg) + m_{\text{rare earth in CRT}} \,(mg)} \times 100\% \qquad (1)$$

After the leaching process, solvent extraction was carried out; di-(2-ethylhexyl) phosphoric acid (DEHPA), tributyl phosphate (TBP), and methyl trioctyl ammonium chloride (Aliquat 336) were employed as extractants, whereas n-hexane and n-heptane were used as diluents. It is important to indicate that a ratio of 1:1 was maintained between the aqueous and the organic phase in all of the assays. The concentration of the extractant (0.5 M, 1.0 M, and 1.5 M) as well as the diluent was assessed in these tests. Finally, in the scrubbing process, a selective removal of europium and yttrium was attempted with HNO_3 and HCl respectively in order to separate the two elements from the organic phase. Yttrium was precipitated from the acidic solution with the addition of oxalic acid, and the result that was precipitated was characterized by X-ray diffraction.

2.3. Methodology for Indium Recovery

The raw material that was used in the project was collected from discarded liquid crystal displays (LCD) of different sizes, brands, and models. The LCDs were dismantled manually; the outer frame, the plastic layer of the LCD monitor, image diffusers, and anti-glare layers were removed until the LCD panel was obtained. After that, a manual size reduction was made to reach a particle size between 0.5–2.0 cm; then, the material was pulverized with a Siebtechnik model T 100 equipment (Siebtechnik, Mülheim an der Ruhr, Germany). The resulted powder was submitted to a granulometric analysis, and three representative sizes of the sample were collected: +298 µm; −149 + 105 µm; and −105 + 74 µm. The powder was characterized by X-ray fluorescence analysis with a S8 Tiger, Bruker equipment (Bruker, Karlsruhe, Germany), and a scanning electron microscopy (SEM)–electric (TESCAN, Brno, Czech Republic) dispersive scanning (EDS) Vega TESCAN. The leaching process was investigated through the use of two acids: HCl and H_2SO_4 with a concentration of 90 g/L. Leaching essays were carried out with magnetic stirring for 24 h. Indium, iron, and aluminum dissolution was monitored by atomic absorption spectroscopy with a Perkin Elmer Analyst 300 spectrometer (Perkin Elmer,

Shelton (CT), USA). The LCD powder was submitted to acid disintegration with HNO_3, HF, and HCl in order to quantify the undissolved elements by atomic absorption. Indium recoveries were obtained by making a mass balance through using Equation (1). For indium recovery form acidic solutions, solvent extraction was carried out. Di-(2-ethylhexyl) phosphoric acid (DEHPA), tributyl phosphate (TBP), and methyl trioctyl ammonium chloride (Aliquat 336) were employed as extractants, whereas n-hexane and diesel were used as diluents. Two ratios of 1:1 and 1:6 between the organic and the aqueous phase respectively were assessed, as well as the concentration of the extractant (0.025–1.000 M). In addition, two organic solvents were tested to determine the optimal conditions to separate indium from iron and aluminum. Once the best extraction results were obtained, the organic phase that had the highest indium load was submitted to re-extraction. The aqueous phase tested was H_2SO_4 at different concentrations (6 M, 8 M, and 10 M), and the amount of indium that was recovered was measured by atomic absorption. Finally, the acid solution was evaporated up to indium precipitated as indium sulfate.

3. Results and Discussion

3.1. Europium and Yttrium Recovery from Cathode Ray Tubes

From the analysis of crystalline phases performed by X-ray diffraction (XRD), it was found that the main constituent compounds of the coating powder were: zinc sulfide (ZnS) and yttrium oxysulfide (Y_2O_2S). During characterization, no europium compounds were found, which was probably due to europium being an activant or dopant of yttrium oxysulfide, but not a differentiable compound. The determination of the elemental chemical composition was carried out by the X-ray analyzer in the scanning electron microscopy (SEM). It should be mentioned that with this analysis, semiquantitative chemical results are obtained, with a limit of detection of 1% (see Figure S1). The results of the elemental chemical composition of the coating powder are shown in Table 1.

Table 1. Elemental chemical composition of coating powder with an X-ray analyzer in SEM.

Metal	Composition (%)
Eu	1.5
Y	8.5
Zn	24.2
Al	9.3
Ti	1.3
Nb	2.6
Pb	7.1
Na	11.9
S	13.7

It was found that zinc is the element that is present in the greatest percentage, followed by sulfur. These results corroborate with the results of the analysis by XRD, in which the zinc sulfide is one of the components of the powder. In the case of rare earths, yttrium has a percentage of 8.5%, coming from the yttrium oxysulfide determined in XRD, whereas europium has a percentage of around 1.5%. In addition, there is a 2.6% of niobium that can arouse interest in their recovery.

In order to extract the rare earths from the CRTs, the powder was used for the extraction of europium and yttrium by acid leaching and solvent extraction. Figure 1 shows the different leaching agents that are used for this purpose. As it can be seen in Figure 1a, the best leaching agent for complex europium and yttrium turned out to be the HNO_3, which was used at a concentration of 200 g/L.

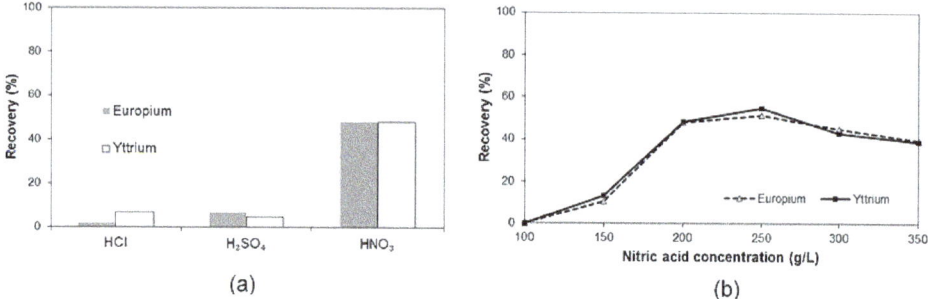

Figure 1. (a) Recovery of europium and yttrium with different leaching agents after 24 h of stirring and 10% of solids; (b) Recovery of europium and yttrium at different concentrations of HNO_3 (10% of solids and 24 h of agitation).

For the leaching carried out with hydrochloric acid, values of recovery were approximately 2% for europium and 7% for yttrium, while with sulfuric acid those values increased to 5% for both metals. Finally, different results are observed in nitric acid leaching tests, in which the increase in recovery is considerable, since approximate values of 48% were reached for both europium and yttrium. In order to determine the optimal concentration of nitric acid, the concentrations were varied from 100 g/L to 350 g/L. The highest recovery of rare earths in solution was obtained when the concentration of HNO_3 was 250 g/L. It should be noted that all of the tests were carried out for 24 h at 10% of solids. It is important to notice that other research studies have demonstrated that sulfuric acid or hydrochloric acid are suitable as well for yttrium dissolution [27]. However, in this case, only nitric acid resulted to be the best leaching agent for rare earths recovery.

Based on these results, tests were carried out to study the influence of leaching time on the recovery of europium and yttrium. In Figure 2a, europium and yttrium recoveries are shown at different leaching times. As the leaching time was increased, higher europium and yttrium recoveries were obtained, reaching the maximum recoveries of both metals at 72 h of agitation. Finally, it can be seen that the curve tends to have an asymptotic shape, so at higher values of leaching time, the stabilization of the curve would be expected in practically constant recovery values.

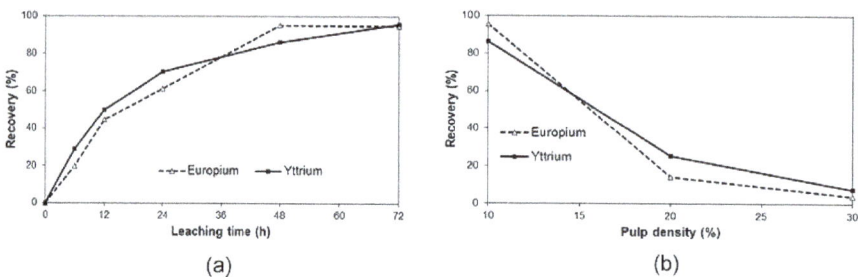

Figure 2. (a) Recovery of europium and yttrium with HNO_3 at 250 g/L and 10% of solids; (b) Recovery of europium and yttrium at different percentages of solids with HNO_3 at 250 g/L for 48 h of agitation.

To complement the study of the leaching process, the influence of pulp density was assessed. In Figure 2b, the europium and yttrium recoveries are shown for leaching tests with different percentages of solids at 250 g/L of HNO_3 for 48 h of stirring. In the case of europium, the recovery decreases from 95% to values of 14% and 4% for a pulp density of 20% and 30%, respectively. On the other hand, yttrium recoveries decrease from 86% to 25% and 8% for pulp densities of 20% and 30%, respectively.

Once the leaching conditions were optimized, the solvent extraction stage was carried out in order to separate the europium and yttrium in the organic phase of the aqueous phase, considering that europium is a light rare earth and yttrium is a heavy rare earth. This classification is according to the atomic masses of the rare earths. In the case of yttrium, there is a phenomenon called lanthanide contraction, which consists in the atomic radius contraction, making this element a heavy rare earth [16]. The n-heptane was used as the organic diluent, and the following substances were used as extractants: di-(2-ethylhexyl) phosphoric acid (DEHPA), tributyl phosphate (TBP), and trioctyl ammonium chloride (Aliquat 336). The ratio between the aqueous phase and the organic phase was 1:1, whereas the contact time between both phases during all of the tests was 10 min. Figure 3 shows the recovery of europium and yttrium in the organic phase.

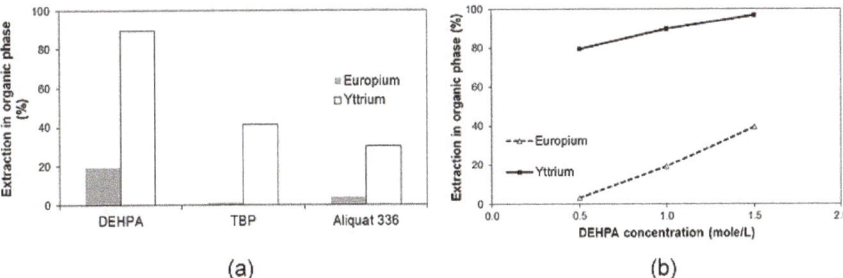

Figure 3. (a) Recovery of europium and yttrium in the organic phase with different extractant agents (extractants concentration in n-heptane: 1 M); (b) Recovery of europium and yttrium at different concentrations of di-(2-ethylhexyl) phosphoric acid (DEHPA).

Figure 3a shows that there is a high preference for yttrium with the three extractants used, with DEHPA being the one that showed the greatest separation factor (8.67) and recovery (90%). In addition, for europium recovery, a separation factor of 0.24 was achieved with DEHPA as the extractant, whereas a separation factor of 0.01 was obtained with TBP. Therefore, any of the tested extractants are suitable for europium recovery from the aquous solution. On the other hand, when observing Figure 3b, an increase in the DEHPA concentration increases the recovery of both elements in the organic phase. For a concentration of 0.5 M, europium recovery was 3% and 80% for yttrium, while for a concnetration 1.5 M, it was 40 and 97%, respectively. The last results are not what is desired, since the aim is to separate both elements in the two phases. For both of the analyzed cases, a concentration of 1 M of DEHPA allowed the balance between a greater extraction of yttrium and a lower co-extraction of europium in the organic phase. A lower concentration of DEHPA implies a lower extraction of europium, while with a higher concentration, there are high extractions of both metals in the organic phase. In addition, although zinc and iron are dissolved in the aqueous solution, they do not affect yttrium extraction at all when DEHPA was used as an extractant. For the entire set of solvent extraction essays, elements such as Zn or Al were not detected in the organic phase after samples analysis via atomic absorbtion. Consequently, the interferences of these elements were considered negligible in this process. This result is in accordance with other works that found DEHPA to be the best substance for yttrium recovery [26]. The results shown in Figure 3a,b suggest that DEHPA is the best extractant for yttrium recovery from the aquous phase, due to the high recovery (90%) and high separation factors (8.67) that this substance presents (see Appendix A, Tables A1–A3).

Figure 4a shows the purification step of the organic phase (Scrubbing) with HNO_3 in order to remove the co-extracted europium. For this, the organic phase was put in contact with an aqueous phase corresponding to solutions of nitric acid at different concentrations.

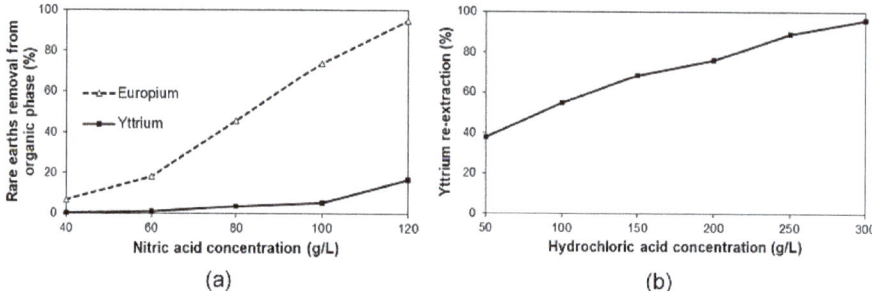

Figure 4. (a) Removal of europium with HNO$_3$; (b) Re-extraction of yttrium with HCl.

When increasing the concentration of HNO$_3$, the removal of europium increases. At a concentration of 120 g/L, a removal of europium of 95% is reached with a loss of yttrium in the organic phase of 17%. Then, the yttrium that was contained in the organic phase, which was previously purified, was re-extracted into a new aqueous phase. So, solutions of different concentrations of hydrochloric acid were used in order to analyze the influence of the concentration of acid in the percentage of yttrium re-extraction. In Figure 4b, it can be observed how the re-extraction of yttrium varies with increasing HCl concentrations from 50 g/L to 300 g/L. Yttrium recovery of 95% is obtained with an HCl concentration of 300 g/L. Lower concentrations of HCl do not allow acceptable yttrium recoveries toward a new aqueous phase.

Finally, the yttrium was precipitated from the aqueous phase with the addition of oxalic acid at a pH of 1.5. The pH was regulated with the addition of sodium hydroxide, and the amount of oxalic acid was four times higher than the stoichiometric value. With these conditions, a 99% precipitation of the yttrium contained in the purified aqueous phase was achieved. Subsequently, the product precipitated in the form of yttrium oxalate was calcined for 2 h at 600 °C to transform it to yttrium oxide with a purity of 68% [29]. The recovery of yttrium oxalate that was acomplished during the precipitation stage is very similar to those that have been carried out in other works [27]. Nevertheless, in contrast with other works, solvent extraction and srcubbing stages were implemented in order to separate yttrium from other impurities. In addition, a calcination step is essential for transforming yttrium oxalate to yttrium oxide, which is a commercial salt. It should be noted that this product can be sold commercially (see Table S1, Figures S2–S4).

In Figure 5, a process flow diagram for yttrium recovery is proposed. The best conditions that were found in this research were employed in the design of the process flow diagram. As other publications, this process incorporates a complete study of all of the parameters related to leaching with acids. Pulp density attracted special attention, since the more CRT powder that is treated, the more profitable the extraction of yttrium. Nevertheless, in the present proposed process, unlike other research studies [27,28], a precipitation step is suggested for yttrium recovery as an oxide, since this salt has a high commercial value.

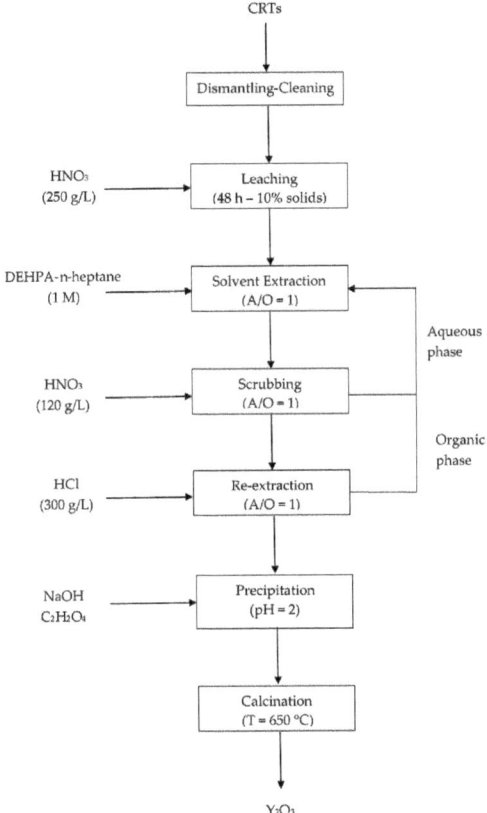

Figure 5. Process flow diagram for yttrium oxide recovery from CRT powder.

3.2. Indium Recovery from LCDs

The elemental chemical composition was analyzed by a scanning electron microscopy (SEM) through an X-ray analyzer. Through this analysis, the presence of indium was verified on the LCD panel. From this analysis, a surface section of the LCD panel was considered and the results obtained from chemical composition are semiquantitative. Characterization carried out by X-ray diffraction (XRD) was impossible, since the diagram obtained shows that the LCD panel is an amorphous material, so a chemical analysis was carried out using X-ray fluorescence (FRX). The chemical characterization of the powder that was carried out by an X-ray fluorescence analysis is shown in Table 2.

Table 2. Elemental chemical composition of the liquid crystal display (LCD) panel using X-ray fluorescence (FRX).

Element	Composition (%)
Si	25.40
Al	7.55
Ca	4.70
Sr	3.70
Mg	0.90
Cl	0.10
Sn	0.09
Fe	0.08
P	0.05
In	0.03

From the results obtained by X-ray fluorescence, the presence of silicon is observed, because the panel is constituted by silicon dioxide (SiO_2), calcium, strontium, and magnesium. These elements are part of the glass, since these they are fluxes that lower the panel melting point. In addition, 0.03% of indium was obtained, which corresponds to 300 mg/g of the panel. For the recovery of indium from the LCD screens, the screens were powdered up to a particle size of 274 µm. The powder obtained was subjected to acid leaching in order to dissolve the indium. For this purpose, the concentration of the leaching agents, the leaching time, and the percentage of solids were varied, as shown in Figure 6.

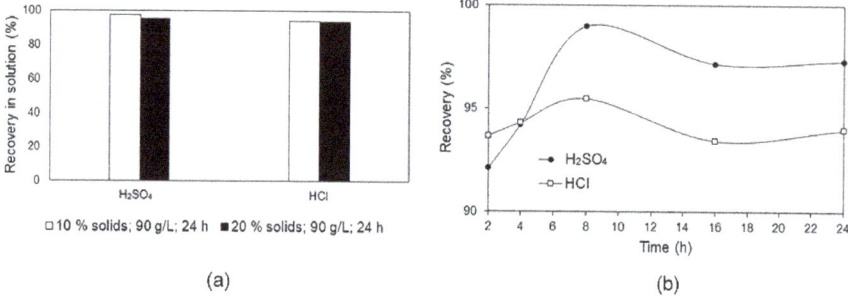

Figure 6. (a) Influence of the percentage of solids in the recovery of indium with H_2SO_4 and HCl; (b) Influence of the leaching time in the recovery of indium with H_2SO_4 and HCl (20% of solids).

In Figure 5a, it is observed that there is not much difference as far as the use of the leaching agent is concerned. However, the use of H_2SO_4 is preferable, since indium sulfate is easier to transform into indium oxide in later stages. Another aspect to keep in mind is that when the percentage of solids varies, the recovery of indium in solution does not fall below 90%. In Figure 6b, an indium recovery of 98% is observed at 8 h when H_2SO_4 is used, while 8 h is required to recover 94% of the indium when HCl is used as a leaching agent. It was also determined that at the time of 2 h of leaching, the recovery of other elements such as iron and aluminum was 24% and 4%, respectively. If the leaching time is increased, the dissolution of these elements also increases, so it is recommended to not exceed the leaching time beyond 4 h.

For the solvent extraction stage, di-(2-ethylhexyl) phosphoric acid (DEHPA) and tributyl phosphate (TBP) were used as extractants, whereas diesel and *n*-hexane were used as diluents. The pH in all of the assays was maintained at 1.7, and the contact time in all of the cases was 10 min. In Figure 7, the results obtained with both diluents are shown.

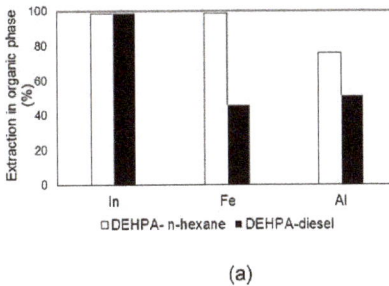 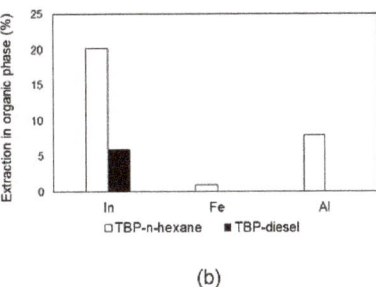

Figure 7. (a) Recovery of In, Fe, and Al with DEHPA (1.0 M, pH = 1.7) with *n*-hexane and diesel as diluents; (b) Recovery of In, Fe and Al with tributyl phosphate (TBP) (1.0 M, pH = 1.7) with *n*-hexane and diesel as diluents.

At a pH value lower than 1.7, a preferred extraction of other metals such as Fe^{2+} and Al^{3+} is observed instead of In^{3+} ions. On the other hand, the high content of H^+ ions present at this pH prevents the In^{3+} ion from being captured [30] and displaces the equilibrium to the left in the extraction reaction of indium, decreasing the extraction of it [31]. Therefore, a pH value of 1.7 was chosen, where a substantial increase in the extraction of indium is observed in the organic phase, reaching an average extraction of 99%. This phenomenon is due to the decrease in the concentration of H^+ ions, with the extraction of indium being favored according to Le Chatelier's principle. Finally, it can be observed that the extraction of Fe^{2+} and Al^{3+} decreases at pH = 1.7, since there is a hydrolysis of the Fe^{2+} and Al^{3+} ions, which makes extraction difficult [30]. From Figure 7a,b, it is determined that the best extractant is di-(2-ethylhexyl) phosphoric acid (DEHPA) with diesel as the diluent, since with the employment of TBP, indium recoveries are lower than those obtained with DEHPA. In addition, iron and aluminum don't affect indium extraction, although they diffuse to the organic phase. In the case of iron, a 46% recovery in the organic phase was obtained. This finding is similar to the results obtained by Yang et al. [26], since they found that 1 M of Cyanex 923 is the extractant that is required for indium extraction with a 47% iron removal. Unlike the group of Yang et al., the selected extractant in this research for indium recovery was DEHPA, which is suitable as well for rare earths extraction.

Figure 8 shows how the extraction of indium, iron, and aluminum with DEHPA at different concentrations occur. As it can be seen in Figure 8a, the recovery of indium is high when the concentration of DEHPA varies from 0.25 mol/L to 1.00 mol/L, regardless of whether diesel or *n*-hexane is used as the diluent. In the case of iron, when the concentration of the extractant is increased to 1.00 mol/L with diesel, the recovery is 50%. When *n*-hexane is used, the recovery of iron did not undergo any variation when increasing the concentration of DEHPA. Finally, in the case of aluminum, it is observed that when the concentration of the extractant increased the recovery of aluminum in the organic phase, it also increased with both diluents. Thus, for the solvent extraction stage, it is recommended to use DEHPA as an extractant with a concentration of 0.25 mol/L with diesel as a diluent, because it is less expensive than *n*-hexane.

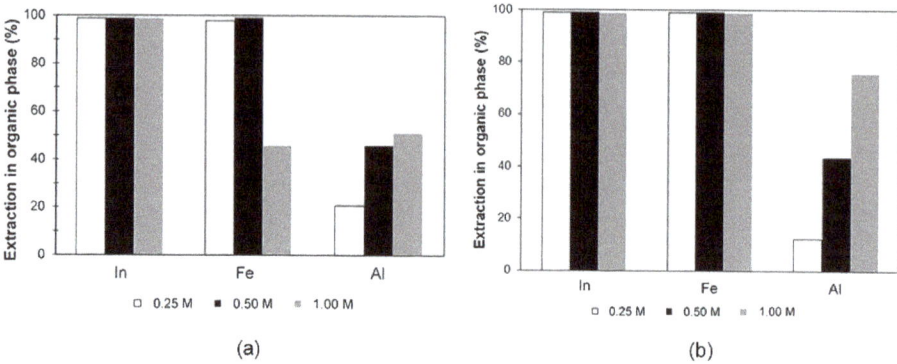

Figure 8. (a) Recovery of In, Fe, and Al with DEHPA and diesel; (b) Recovery of In, Fe, and Al with DEHPA and *n*-hexane.

For the re-extraction stage, the organic phase was diluted to allow a better diffusion of the indium to the aqueous phase constituted by the concentrated H_2SO_4 solution. Figure 9 shows the results of re-extraction at different H_2SO_4 concentrations.

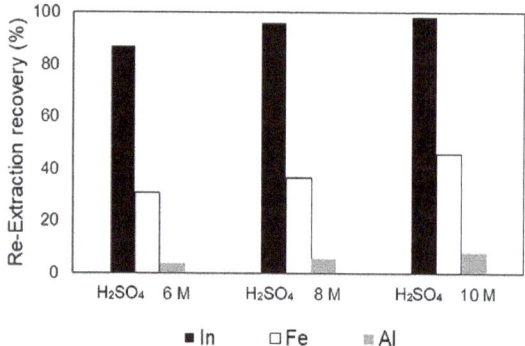

Figure 9. Influence of the concentration of H_2SO_4 on the re-extraction of indium.

The ratio between the aqueous phase and organic phase was six, and the contact time between both phases was 5 min. An increase in the recovery of all of the metals is observed when increasing the H_2SO_4 concentration from 6 mol/L to 10 mol/L. Therefore, a concentration of 8 mol/L of H_2SO_4 was chosen to obtain a recovery of 95% of indium, diminishing the re-extractions of iron and aluminium to 37% and 6%, respectively. Afterwards, the solution was evaporated in order to precipitate the indium sulfate. The precipitated solid was subjected to calcination for 2 h at 700 °C to obtain indium oxide (see Figure S5). In Figure 10, a process flow diagram is shown for indium oxide recovery from LCD panels. Similar to other research studies, this process incorporates the solvent extraction step, which is essential for indium separation from LCD impurities such as iron and aluminum. In contrast with other works [26], this process combines a solvent extraction step and a precipitation for the recovery of indium oxide, which is a salt of great commercial value.

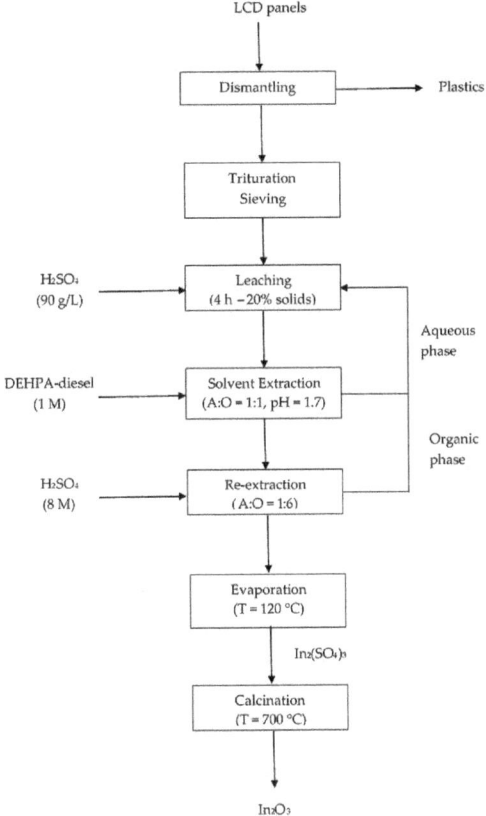

Figure 10. Process flow diagram for indium oxide recovery from LCD panels.

4. Conclusions

The recovery of rare earths from WEEE is feasible by the hydrometallurgical route, using for this purpose processes such as: acid leaching, solvent extraction, and precipitation. It should be noted that the WEEE from which it is desired to extract the rare earths must be pulverized in order to release the elements of interest.

In the case of europium and yttrium dissolution from the CTRs, the best leaching agent turned out to be HNO_3 at 250 g/L of concentration. Leaching tests demonstrated that leaching times longer than 24 h allow higher recoveries, because there is a longer contact time between the nitric acid and the components of the coating powder that may be dissolved. Similarly, higher amounts of other metals such as zinc make it necessary to increase the leaching time to dissolve rare earths. In addition, when pulp density increases, the solution is saturated with the formed europium and yttrium nitrates. On the other hand, the reduction of the acid/powder ratio can also negatively influence rare earths dissolution, since there is less formation of soluble products, and therefore lower recoveries.

The solvent extraction stage was carried out with DEHPA at a concentration of 0.5 mol/L with a contact time of 5 min and using n-heptane as a diluent. Subsequently, a purification step of the organic phase was carried out with HNO_3 of 120 g/L concentration to extract the europium. Finally, the re-extraction of yttrium was carried out in the aqueous phase using HCl of 300 g/L concentration. The yttrium precipitation was carried out with four times the stoichiometric amount of oxalic acid, and the solid obtained was calcined at 600 °C for 2 h to obtain yttrium oxide of 68% purity.

For the extraction of indium from the LCD screens, H_2SO_4 leaching agent was used at a concentration of 90 g/L at 20% solids for 4 h. The extraction by solvents was carried out with the DEHPA extractant at a concentration of 0.50 mol/L with diesel as the diluent. The contact time was 10 min, and the aqueous phase–organic phase ratio was 1. For the re-extraction, H_2SO_4 was used at a concentration of 10 mol/L with a contact time of 5 min. Finally, when the solution evaporated, the precipitated indium was obtained in the form of sulfate, which was calcined at 700 °C for 2 h to obtain indium oxide.

Supplementary Materials: The following are available online at http://www.mdpi.com/2075-4701/8/10/777/s1, Figure S1: CRT powder SEM image, Figure S2: Yttrium oxide SEM image, Figure S3: Yttrium oxide chemical characterization with SEM, Figure S4: Yttrium oxide diffractogram obtained after calcination, Figure S5: Indium oxide diffractogram obtained after calcination, Table S1: Yttrium oxide characterization.

Author Contributions: Conceptualization, E.T.; Data curation, S.G.; Formal analysis, S.G.; Investigation, C.R. and E.V.; Methodology, C.R. and E.V.; Project administration, E.T.; Writing—original draft, S.G.; Writing—review & editing, S.G.

Funding: This research was funded by the Project PIS 039 12 of the Escuela Politécnica Nacional, which was executed in the Department of Extractive Metallurgy.

Acknowledgments: The authors address their thanks to the Escuela Politécnica Nacional for the support in the development of the present research.

Conflicts of Interest: The authors declare no conflict of interest. The funders had no role in the design of the study; in the collection, analyses, or interpretation of data; in the writing of the manuscript, or in the decision to publish the results.

Appendix A

Appendix A.1. Analysis of Europium and Yttrium by Disintegration in a Microwave Oven

Disintegration in a microwave oven is a technique that allows the dissolution, in an aqueous phase, of a solid sample in order to determine the concentration of the metals present in the sample by means of the atomic absorption technique. To carry out the disintegration, a Samsung microwave oven was used. He procedure followed for the disintegration of the tailings obtained in the leaching process is detailed [32]:

1. The tailings were dried in a stove, at 110 °C for 24 h.
2. 0.1 g of the dry tailings was weighed and placed in a Teflon reactor.
3. 3 mL of HNO_3 and 3 mL of HF were added.
4. The reactor was sealed and it was taken to the microwave oven where it was left for 2.5 min at medium power.
5. The reactor was cooled to 30 min.
6. The reactor was opened and 5 mL of HCl was added.
7. The reactor was sealed and it was taken to the microwave oven for a time of 2.5 min at medium power.
8. The reactor was cooled to 30 min.
9. The content of the reactor was placed in a flask and adjusted to a volume of 50 mL

The solution obtained by disintegration of the tailings was analyzed by atomic absorption spectrometry to determine the concentration of europium and yttrium in solution, with these data the respective metallurgical balance was completed in each of the tests carried out.

Appendix A.2. Analysis of Indium by Disintegration in a Microwave Oven

In this case, 100 mg of sample were weighed and put in a closed Teflon reactor. Then 3 mL of HNO_3 and 3 mL of HF were added to the sample and the mixture was taken to a Samsung microwave oven with a power of 50 W in a lapse of 2.5 min. Afterwards the sample was cooled for 40 min. Then,

5 mL of HCl was added to the reactor and introduced into the microwave for 2.5 min and power of 50 W. Once the process was completed, the solution was adjusted to 100 mL. Finally, the disaggregated solutions to be analyzed were sent by atomic absorption spectrophotometry.

Appendix A.3. Examples of Metallurgical Balance in Solvent Extraction Process

Table A1. Europium and yttrium recovery and separation factor with DEHPA

Conditions				
Extractant		DEHPA		
Concentration		1 M		
Diluent		n-heptane		
Ratio A/O		1		
Contact time		10 min		
Mass Balance in Solvent Extraction				
	Volume	Y (mg/L)	Y (mg)	Distribution (%)
Aqueous phase (mL)	15.00	910.00	13.65	10.34
Organic phase (mL)	15.00	7 890.00	118.35	89.66
			132.00	100.00
Separation factor of Y		8.67		
Recovery of Y		89.66%		
	Volume	Eu (mg/L)	Eu (mg)	Distribution (%)
Aqueous phase (mL)	15.00	481.00	7.21	80.70
Organic phase (mL)	15.00	115.00	1.72	19.30
			8.94	100.00
Separation factor of Eu		0.24		
Recovery of Eu		19.30%		

Table A2. Europium and yttrium recovery and separation factor with TBP.

Conditions				
Extractant		TBP		
Concentration		1 M		
Diluent		n-heptane		
Ratio A/O		1		
Contact time		10 min		
Mass Balance in Solvent Extraction				
	Volume	Y (mg/L)	Y (mg)	Distribution (%)
Aqueous phase (mL)	15.00	5 170.00	77.55	58.75
Organic phase (mL)	15.00	3 630.00	54.45	41.25
			132.00	100.00
Separation factor of Y		0.70		
Recovery of Y		41.25%		
	Volume	Eu (mg/L)	Eu (mg)	Distribution (%)
Aqueous phase (mL)	15.00	588.00	8.82	98.66
Organic phase (mL)	15.00	8.00	0.12	1.34
			8.94	100.00
Separation factor of Eu		0.01		
Recovery of Eu		1.34%		

Table A3. Europium and yttrium recovery and separation factor with Aliquat 336.

Conditions	
Extractant	Aliquat 336
Concentration	1 M
Diluent	n-heptane
Ratio A/O	1
Contact time:	10 min

Mass Balance in Solvent Extraction				
	Volume	Y (mg/L)	Y (mg)	Distribution (%)
Aqueous phase (mL)	15.00	6 070.00	91.05	69.93
Organic phase (mL)	15.00	2 610.00	39.15	30.07
			130.20	100.00
Separation factor of Y			0.43	
Recovery of Y			30.07%	
	Volume	Eu (mg/L)	Eu (mg)	Distribution (%)
Aqueous phase (mL)	15.00	530.00	7.95	96.01
Organic phase (mL)	15.00	22.00	0.33	3.99
			8.28	100.00
Separation factor of Eu			0.04	
Recovery of Eu			3.99%	

References

1. Kahhat, R.; Kim, J.; Xu, M.; Allenby, B.; Williams, E.; Zhang, P. Exploring e-waste management systems in the United States. *Resour. Conserv. Recycl.* **2008**, *52*, 955–964. [CrossRef]
2. Ongondo, F.O.; Williams, I.D.; Cherrett, T.J. How are WEEE doing? A global review of the management of electrical and electronic wastes. *Waste Manag.* **2011**, *31*, 714–730. [CrossRef] [PubMed]
3. Babu, B.R.; Parande, A.K.; Basha, C.A. Electrical and electronic waste: A global environmental problem. *Waste Manag. Res.* **2007**, *25*, 307–318.
4. Barba, Y.; Adenso, B.; Hopp, M. An analysis of some environmental consequences of European electrical and electronic waste regulation. *Resour. Conserv. Recycl.* **2008**, *52*, 481–495. [CrossRef]
5. Constantino, F.; De Minicis, M.; Di Gravio, G. Analytic network process for WEEE reverse logistics networks. In Proceedings of the 13th International Symposium on Logistics, Bagkok, Thailand, 6–8 July 2008; Centre for Concurrent Enterprise: Nottingham, UK, 2008.
6. Herat, S. Recycling of cathode ray tubes (CRTs) in electronic waste. *Clean-Soil Air Water* **2008**, *36*, 19–24. [CrossRef]
7. Cui, J.; Zhang, L. Metallurgical recovery of metals from electronic waste: A review. *J. Hazard. Mater.* **2008**, *158*, 228–256. [CrossRef] [PubMed]
8. Oguchi, M.; Sakanakura, H.; Terazono, A.; Takigami, H. Fate of metals contained in waste electrical and electronic equipment in a municipal waste treatment process. *Waste Manag.* **2012**, *32*, 92–103. [CrossRef] [PubMed]
9. Méar, F.; Yot, P.; Cambon, M.; Ribes, M. The characterization of waste cathode-ray tube glass. *Waste Manag.* **2006**, *26*, 1468–1476. [CrossRef] [PubMed]
10. Romero, J.L.; McCord, S.A. Rare Earth Elements: Procurement, Application, and Reclamation. Available online: https://prod.sandia.gov/techlib-noauth/access-control.cgi/2012/126316.pdf (accessed on 8 September 2018).
11. Henderson, P. *Rare Earth Element Geochemistry*, 3rd ed.; Elsevier: London, UK, 1984; pp. 380–410.
12. Sinha, S. *Complexes of the Rare Earths*, 4th ed.; Elsevier: Geneva, Switzerland, 1966; pp. 17–81.
13. Humphries, M. *Rare Earth Elements: The Global Supply Chain*; CRS Report for Congress: Washington, DC, USA, 2013.
14. Golev, A.; Scott, M.; Erskine, P.D.; Ali, S.H.; Ballantyne, G.R. Rare earths supply chains: Current status, constraints and opportunities. *Resour. Policy* **2014**, *41*, 52–59. [CrossRef]

15. Jordens, A.; Cheng, Y.P.; Waters, K.E. A review of the beneficiation of rare earth element bearing minerals. *Miner. Eng.* **2013**, *41*, 97–114. [CrossRef]
16. Gupta, C.; Krishnamurthy, N. *Extractive Metallurgy of Rare Earths*, 2nd ed.; CRC PRESS: Boca Raton, FL, USA, 2015; pp. 280–340.
17. Zepf, V. An Overview of the Usefulness and Strategic Value of Rare Earth Metals. In *Rare Earths Industry*, 1st ed.; Borges de Lima, I., Leal Filho, W., Eds.; Elsevier Inc.: Boston, MA, USA, 2016; pp. 3–17. ISBN 978-0-12-802328-0.
18. Peelman, S.; Sun, Z.H.I.; Sietsma, J.; Yang, Y. Leaching of Rare Earth Elements. In *Rare Earths Industry*, 1st ed.; Borges de Lima, I., Leal Filho, W., Eds.; Elsevier: Boston, MA, USA, 2015; pp. 319–334. ISBN 978-0-12-802328-0.
19. Alex, P.; Suri, A.K.; Gupta, C.K. Processing of xenotime concentrate. *Hydrometallurgy* **2001**, *50*, 331–338. [CrossRef]
20. Zhuang, X.; He, W.; Li, G.; Huang, J.; Ye, Y. Materials separation from waste liquid crystal displays using combined physical methods. *Pol. J. Environ. Stud.* **2012**, *21*, 1921–1927.
21. Yang, J. Recovery of Indium from End-of-Life Liquid Crystal Displays. Bachelor's Thesis, Thesis for the Degree of Licentiate of Engineering. Chalmers University of Technology, Gothenburg, Sweden, May 2012.
22. Cook, N.J.; Ciobanu, C.L.; Williams, T. The mineralogy and mineral chemistry of indium in sulphide deposits and implications for mineral processing. *Hydrometallurgy* **2011**, *108*, 226–228. [CrossRef]
23. Tolcin, A.C. Mineral Commodities Summary 2016: Indium. *US Geol. Surv.* **2016**, *703*, 80–81.
24. Koleini, S.M.J.; Mehrpouya, H.; Saberyan, K.; Abdolahi, M. Extraction of indium from zinc plant residues. *Miner. Eng.* **2010**, *23*, 51–53. [CrossRef]
25. Alfantazi, A.M.; Moskalyk, R.R. Processing of indium: A review. *Miner. Eng.* **2003**, *16*, 687–694. [CrossRef]
26. Yang, J.; Retegan, T.; Steenari, B.; Ekberg, C. Recovery of indium and yttrium from Flat Panel Display waste using solvent extraction. *Sep. Purif. Technol.* **2016**, *166*, 117–124. [CrossRef]
27. De Michelis, I.; Ferella, F.; Varelli, E.; Vegliò, F. Treatment of exhaust fluorescent lamps to recover yttrium: Experimental and process analyses. *Waste Manag.* **2011**, *31*, 2559–2568. [CrossRef] [PubMed]
28. Innocenzi, V.; De Michelis, I.; Ferella, F.; Vegliò, F. Recovery of yttrium from cathode ray tubes and lamps' fluorescent powders: Experimental results and economic simulation. *Waste Manag.* **2013**, *33*, 2390–2396. [CrossRef] [PubMed]
29. Jorjani, E.; Shahbazi, M. The production of rare earth elements group via tributyl phosphate extraction and precipitation stripping using oxalic acid. *Arabian J. Chem.* **2016**, *9*, S1532–S1539. [CrossRef]
30. Li, X.; Deng, Z.; Li, C.; Wei, C.; Li, M.; Fan, G.; Rong, H. Direct solvent extraction of indium from a zinc residue reductive leach solution by D_2EHPA. *Hydrometallurgy* **2015**, *156*, 1–5. [CrossRef]
31. Ritcey, M.; Ashbrook, W. *Solvent Extraction: Principles and Applications to Process Metallurgy*, 1st ed.; Elsevier: Michigan, IN, USA, 1984; pp. 1–362. ISBN 0444417516.
32. Morales, D. Procesamiento de un Mineral Aurifero Refractario para la Recuperacion de oro. Bachelor's Thesis, Thesis for the Degree of Licentiate of Chemical Engineering. Escuela Politécnica Nacional, Quito, Ecuador, 10 February 2011.

© 2018 by the authors. Licensee MDPI, Basel, Switzerland. This article is an open access article distributed under the terms and conditions of the Creative Commons Attribution (CC BY) license (http://creativecommons.org/licenses/by/4.0/).

MDPI
St. Alban-Anlage 66
4052 Basel
Switzerland
Tel. +41 61 683 77 34
Fax +41 61 302 89 18
www.mdpi.com

Metals Editorial Office
E-mail: metals@mdpi.com
www.mdpi.com/journal/metals

www.ingramcontent.com/pod-product-compliance
Lightning Source LLC
LaVergne TN
LVHW070701100526
838202LV00013B/1010